SpringerBriefs in Geoethics

Editor-in-Chief

Silvia Peppoloni, National Institute of Geophysics and Volcanology (INVG), Rome, Italy

Series Editors

Nic Bilham, University of Exeter, Penryn, UK

Peter T. Bobrowsky, Geological Survey of Canada, Sidney, Canada

Vincent S. Cronin, Baylor University, Waco, USA

Giuseppe Di Capua ⓘ, National Institute of Geophysics and Volcanology (INVG), Rome, Italy

Iain Stewart, University of Plymouth, Plymouth, UK

Artur Sá, University of Trás-os-Montes and Alto Douro, Vila Real, Portugal

Rika Preiser, Stellenbosch University, Stellenbosch, South Africa

SpringerBriefs in Geoethics envisions a series of short publications that aim to discuss ethical, social, and cultural implications of geosciences knowledge, education, research, practice and communication. The series SpringerBriefs in Geoethics is sponsored by the IAPG—International Association for Promoting Geoethics (http://www.geoethics.org).

The intention is to present concise summaries of cutting-edge theoretical aspects, research, practical applications, as well as case-studies across a wide spectrum.

SpringerBriefs in Geoethics are seen as complementing monographs and journal articles, or developing innovative perspectives with compact volumes of 50 to 125 pages, covering a wide range of contents comprising philosophy of geosciences and history of geosciences thinking; research integrity and professionalism in geosciences; working climate issues and related aspects; geoethics in georisks and disaster risk reduction; responsible georesources management; ethical and social aspects in geoeducation and geosciences communication; geoethics applied to different geoscience fields including economic geology, paleontology, forensic geology and medical geology; ethical and societal relevance of geoheritage and geodiversity; sociological aspects in geosciences and geosciences-society-policy interface; geosciences for sustainable and responsible development; geoethical implications in global and local changes of socio-ecological systems; ethics in geoengineering; ethical issues in climate change and ocean science studies; ethical implications in geosciences data life cycle and big data; ethical and social matters in the international geoscience cooperation.

Typical topics might include:

- Presentations of core concepts
- Timely reports on state-of-the art
- Bridges between new research results and contextual literature reviews
- Innovative and original perspectives
- Snapshots of hot or emerging topics
- In-depth case studies or examples

All projects will be submitted to the series-editor for consideration and editorial review.

Each volume is copyrighted in the name of the authors. The authors and IAPG retain the right to post a pre-publication version on their respective websites.

The Series in Geoethics is initiated and supervised by Silvia Peppoloni and an editorial board formed by Nic Bilham, Peter T. Bobrowsky, Vincent S. Cronin, Giuseppe Di Capua, Rika Preiser, Artur Agostinho de Abreu e Sá, Iain Stewart.

More information about this series at https://link.springer.com/bookseries/16482

Sandra Paula Villacorta Chambi
Editor

Geoethics in Peru

A Pathway for Latin America

 Springer

Editor
Sandra Paula Villacorta Chambi
Charles Darwin University
Sydney, Australia

ISSN 2662-6780 ISSN 2662-6799 (electronic)
SpringerBriefs in Geoethics
ISBN 978-3-030-86730-0 ISBN 978-3-030-86731-7 (eBook)
https://doi.org/10.1007/978-3-030-86731-7

This Springer imprint is published by the registered company Springer Nature Switzerland AG
The registered company address is: Gewerbestrasse 11, 6330 Cham, Switzerland

To our families, for their support and patience

To the members of IAPG Peru for a new future for the Peruvian society

Foreword by Silvia Peppoloni

In 2014, the International Association for Promoting Geoethics (IAPG) established a section in Peru. Thanks to the activism and enthusiasm of a part of the young and vibrant Peruvian geoscientific community, geoethics has found a fertile place in Peru to take root and develop. The national geoethics community, consolidated around the IAPG-Peru, has increased in number over time and its impact has grown with tangible initiatives addressed to the local scientific community and the whole Peruvian society. For example, MinerLima, the International exhibition on minerals, organized since 2015 by IAPG-Peru, for promoting geoscience knowledge and geoscientific culture, to strengthen the social role of geoscientists which is fostering a more sustainable development of Peruvian society. Today, the IAPG-Peru is a concrete reality of the Peruvian scientific and cultural landscape, an inclusive multidisciplinary meeting point, open to dialogue with local stakeholders.

This is the second book published in the SpringerBriefs in Geoethics and it is significant that it is written by colleagues from Peru, one of the economically emerging Latin American countries.

South and Central Americas are extraordinary socio-cultural and environmental laboratories, in which the contaminations between Western cultural traditions and local millenary cultures find a complex synthesis characterized by lights and shadows.

In Peru, the relationship that binds human communities to their territories is characterized by strong symbolism meanings, arising from the concept of Pachamama (Mother Earth) and materializing in feelings and actions to protect Nature, as well as by elements of contrast typical of contemporary societies, in which Nature is subject to heavy exploitation and destruction of its ecosystems.

In this ambivalence, Latin America lives all its contradictions, increased by strong social inequalities, by the continuous threat to people's fundamental human rights and by the progressive erosion of the living spaces of local indigenous communities. On the other hand, it is evident the potential of this continent for undertaking innovative and inclusive paths to build new models of sustainable economic development and human progress.

From this perspective, geoethics becomes a proposal capable of reconciling different positions and channelling different interests, expectations and perspectives on important issues such as the defence from natural risks, responsible management of georesources, protection and conservation of the geological heritage and geodiversity, as well as the geological literacy of the population. These are precisely the themes dealt with in the book, accompanied by a reflection from the perspective of geoethical thinking.

For all these reasons, the book constitutes a point of reference in the development of geoethics in Latin America: on the one hand, it aims to be an analysis and a reflection after years of activity of the IAPG-Peru at the service of Peruvian society, on the other hand, it pushes other communities of the continent to undertake a shared path to propose the principles and values of geoethics as a response to the great environmental and social issues of our times.

Silvia Peppoloni
Giuseppe Di Capua
International Association for Promoting Geoethics
Istituto Nazionale di Geofisica e Vulcanologia
Rome, Italy

Foreword by Roberto Greco

In my view, there are several reasons to read Geoethics in Peru.

This book tells us how geosciences began their development in Peru, as well as describing the current complex reality and the many institutions and initiatives that lead this science field. This content alone should be of interest to many academic specialists from many fields such as the history of science, ethics, geosciences, social sciences, education, or natural heritage.

But the goal of this book has a more practical and direct purpose: to interact and mobilize the geoscience community.

By presenting the history of the Peruvian geoscience community, the institutional partners and the relevance of geoscience for society, contribute to give a whole picture of geosciences in Peru, providing a fundamental framework for every professional and academic geoscientist. For geology students, this book offers a few hours of reading the information that usually people need years to collect by readings, participating in scientific events and talking with senior geoscientists.

That for sure will contribute to strengthening the position within the Peruvian geoscience's community.

The special and innovative perspective of geoethics is used to carry a clear message to the community of geoscientists, the need to "humanize" the practice of geosciences. Many arguments and examples support this statement.

Which are the values that sustain geoscientists in their professionals' activities? This is not a trivial question in a field that brings geoscientists to interact with different stakeholders from urban or rural citizens and public administrators to a multinational private company.

This is a needed reflection that should start from the undergraduate course and accompany the professionals along with their life.

There are many suggestions for innovative science and technology policy along with the book that could be useful for the agenda of public administrators not only in Peru.

It is the ambition that the endeavour to create an integrated picture of the geosciences at a national level will soon be followed by other countries as it prepares the geosciences sector for rethinking the social role and geosciences culture inside the society.

This is an original contribution from Latin America for a change of perspective inside the global geoscience's community.

Roberto Greco
Chair of the International Geoscience Education Organization,
Coordinator for Latin America of the International Association
for Promoting Geoethics, Member of the IUGS Commission on
Geoscience Education
Department of Science and Technology Policy
Institute of Geoscience, Unicamp
Campinas, Brazil
greco@unicamp.br

Acknowledgments

Writing a book requires the cooperation of many individual professionals, and coordinating with them has been a big challenge during the COVID-19 pandemic. I value their hard work in the education sector as well as their friendship. My sincere appreciation goes to Silvia Peppoloni and Giuseppe Di Capua, who prepared an in-depth foreword. Their critical comments and thoughtful input helped guide our journey promoting geoethics in Peru and strengthening our efforts. I was also privileged to have the close collaboration of Roberto Greco, the president of the International Geoscience Education Organization. He collaborated in writing a foreword for the book and the chapter on geoscience education. His expertise, insights, and writing skills greatly helped in writing the chapter. I also appreciate the participation of some outstanding professionals from IAPG-Peru, with their contributions to different chapters of the book: Cesar Chacaltana, Pedro Isique, Esteban Manrique, Julio Cardenas, Luis Araujo, and Carlos Toledo.

Giuseppe Di Capua deserves special recognition for the review of the complete document to improve its quality, and for his constant support.

I also would like to thank the team of professionals of WOMEESA: Heather Handley, Johanna Parr, and Anna Petts that made a significant contribution in reviewing the chapters on Geoheritage, mining activity and Palaeontology, respectively, also, my recognition to Chris King for his support reviewing the chapter on geoscience education.

I wish also to thank Juan Carlos Piscoya and Efrain Patiño from the IAPG-Peru Board. I sincerely appreciate their comments on the introduction and the mining activity chapter, respectively.

Finally, I want to recognize my friends: Henry Attfield, Grant Harcourt, and Walter Barnett for checking the grammar of the book.

Sandra Paula Villacorta Chambi

Contents

Contributors

Luis Araujo Ramos Cesar Vallejo University, Trujillo, Peru;
Cesar Vallejo University, Lima, Peru

Julio Cardenas Manzaneda Instituto de Geociencias y Medio Ambiente, Lima, Peru

Cesar Chacaltana Budiel Antonio Ruiz de Montoya University, Pueblo Libre, Peru;
Instituto Geológico Minero Y Metalúrgico, Lima, Peru

Roberto Greco Campinas University, Campinas, Brasil

Pedro Isique Chaname PIASA Consultores SA, Lima, Perú

Esteban Manrique Zuñiga National University of Engineering, Rímac, Peru

Carlos Toledo Gutierrez Antonio Ruiz de Montoya University, Pueblo Libre, Peru;
Charles Darwin University, Casuarina, Australia;
Antonio Ruiz de Montoya University, Lima, Peru

Sandra Paula Villacorta Chambi Charles Darwin University, Casuarina, Australia;
Charles Darwin University, Darwin, Australia

Abbreviations and Acronyms

APCSG	Peruvian Deputy for Social Conflict Prevention and Governance
AUTOCOLCA	Autonomous Authority of Colca and Annexes
CCGN	National Geological Chart Commission
CENEPRED	Peruvian Centre for Estimation, Prevention and Reduction of Disaster Risk
CINGEO	Peruvian Geotechnical Research Community
CISMID	Japanese-Peruvian Centre for Seismic Research and Disaster Mitigation
CONCYTEC	National Council for Science, Technology and Innovation
CONI	Ministry and the National Research Council
CONICET	Argentinian Council for Scientific and Technical Research
CVZ	central volcanic zone of the Andes
ENSO	El Niño Southern Oscillation
GEMMA	Group of Standards for Mass Movements
IAPG	International Association for the Promotion of Geoethics
IGP	Geophysical Institute of Peru
INAAC	National Institute of Concessions and Mining Cadastre
INCITEMI	Scientific and Technological Institute in Mining
INGEMMET	Geological Mining and Metallurgical Institute
INGEOMIN	Institute of Geology and Mining
INIFM	National Institute for Mining Research and Development
MEM	Ministry of Energy and Mines
MinerLima	Lima's International Mineral Show
NPO	Non-profit organization
OVI	Volcanological Observatory of INGEMMET
PUCP	Pontifical Catholic University of Peru
SDG	Sustainable Development Goals
SEGCGS	Specialized Section in Geoethics, Geological Culture and Society
SGL	Geographical Society of Lima
SGP	Geological Society of Peru
SUNARP	Superintendency of Public Registry

UN	United Nations
UNAM	National Autonomous University of Mexico
UNAP	National University of the Altiplano
UNC	National University of Cajamarca
UNDAC	Daniel Alcides Carrión National University
UNESCO	United Nations Educational, Scientific and Cultural Organization
UNI	National University of Engineering
UNJBG	Jorge Basadre Grohmann's National University
UNMSM	San Marcos's National University
UNP	National University of Piura
UNSA	National University of San Agustin
UNSAAC	San Antonio Abad del Cusco National University
UPN	Private University of the North

List of Figures

Geoscience Education in Peru

Conclusions: Future Perspectives of Geoethics in Peru

List of Tables

Introduction

Sandra Paula Villacorta Chambi, Carlos Toledo Gutierrez, and Luis Araujo Ramos

1 Introduction

For those not familiar with Peru; it is a South American country that borders the Pacific Ocean and reaches the top of the Andes mountains and is largely covered by the Amazon Rain forest. Peru is home to more than 33 million people, with a density of 25 inhabitants per square kilometre, one of the lowest in the world. Today, the population is concentrated in a few big cities while the vastness of the bulk country is occupied by less than one million people approximately (INEI, 2009a cited by Guadalupe et al., 2020), living in settlements of a few dozen or a few hundred people.

The current situation in Peru is similar to that in Latin American countries, where the practice of geoethics (Peppoloni & Di Capua, 2015) is still in its beginnings. In Peru, some factors have an enormous influence on a society that suffers an uncertain and conflicting political situation. Peru bases its primary economy on the export (60% to the total amount) of raw materials, and mining represents the first industry of the country (Ministry of Energy and Mines, 2020).

Electronic supplementary material The online version of this chapter (https://doi.org/10.1007/978-3-030-86731-7_1) contains supplementary material, which is available to authorized users.

S. P. Villacorta Chambi (✉)
Charles Darwin University, Casuarina, Australia
e-mail: sandra.villacorta-chambi@cdu.edu.au

C. Toledo Gutierrez
Antonio Ruiz de Montoya University, Pueblo Libre, Peru
e-mail: carlos.toledo@uarm.pe

L. Araujo Ramos
Cesar Vallejo University, Trujillo, Peru

In developing countries like Peru, it is highly relevant to promote sustainable development, which implies working within the framework of the 2030 Agenda for Sustainable Development Goals—SDGs (UN, 2015). It involves efforts to get an adequate educational budget (Crespo et al., 2017; Vasconcelos et al., 2018) in order to achieve the goal 4 "Quality Education" (https://sdgs.un.org/goals, accessed 10 May 2021). Students should acquire knowledge, principles and fundamental values related to sustainability in the present and the future (UN, 2015). In this context, geoethics is a relevant opportunity to support a societal cultural change capable of creating conditions favourable to achieve the SDGs and providing more holistic solutions to problems affecting the interaction between humans and the Earth system (Bobrowsky et al., 2018; Mogk et al., 2018).

This chapter presents an overview of the ongoing programs and projects of the Peruvian Section of the International Association for Promoting Geoethics (IAPG-Peru) to contribute to achieving and strengthening the IAPG goals in the country. Those programs and projects are developed through strategies elaborated from inter- and multidisciplinary perspectives, based on geoscientific knowledge and research, and incorporating local experiences. The organization seeks not only to spread geoscience and geoethical values but also to make citizens aware of their contribution to social well-being. In political, social and economic circumstances like those Peru is facing, professionals and academics are not only discovering the need to reflect deeply but also to focus on the population awareness on topics like environmental protection, sustainable development, inclusiveness policies, reduction of inequalities and construction of a knowledgeable society. This is an immense challenge and all geoscientists are in the front line of it (Peppoloni et al., 2017), being conscious that the social role they can play for the benefit of society can strengthen by adhering to the "Geoethical Promise" (Matteucci et al., 2014).

I promise......

I will practice Geosciences being fully aware of the social implications, and I will do my best to protect the Earth system for the benefit of mankind. ...

I understand my responsibilities towards society, future generations and the Earth for sustainable development. ...

I will put the interest of society foremost in my work. ...

I will never misuse my knowledge of Geosciences, resisting constraint or coercion. ...

I will always be ready to provide my professional assistance when needed, and will be impartial in making my experience available to decision makers. ...

I will continue lifelong development of my geoscientific knowledge. ...

I will always maintain intellectual honesty in my work, being aware of the limits of my ompetencies and skills. ...

I will act to foster progress in the geosciences, the sharing of geoscientific knowledge and the dissemination of the geoethical approach

I will always be fully respectful of Earth processes in my work as a geoscientist. I promise!

2   The Establishment of the IAPG Peruvian Section

IAPG Peru was the first association dedicated exclusively to the promotion and study of geoethics in Peru, and its members were responsible for much of the significant progress of this science in Latin America. Its distinctive character as a centre for discussion and research on ethics in geosciences in Peru was established by its founders: Sandra Villacorta, Cosme Perez-Puig, Gregorio Villacorta and Juan Carlos Piscoya in 2014. Since that year, its activities have been directed by a board and its policies are largely shaped by the IAPG Executive Council.

The Peruvian section began as a small group of professionals in Lima, but quickly grew into a scientific association with a nationwide membership. IAPG Peru development into a large independent section of IAPG is the result of constant work of Peruvian committee evaluated by the international board and its vision of the importance of applying an ambitious cooperative dissemination plan at all levels. This vision allowed the section to establish independence and become more active in research activities. Participation of philosophers with geoscientists has been a good joint venture. Both of them first embraced the geoethical code, the Cape Town Statement on Geoethics (Di Capua et al., 2017), then promoted its endorsement by Peruvian entities.

The main achievements from the activities by the Peruvian section are the positive impacts that have carried over the Peruvian geoscientist community. The coordination of major events, collaborative projects and initiatives and signed agreements are summarized below. These initiatives include the publication of an open letter from IAPG Peru on Risk Management in Peru in 2017, the creation of the Specialized Section in Geoethics, Geological Culture and Society (SEGCGS), registration in the Public Registries of Peru and relations with student associations.

3   Projects and Actions to Promote Geoethics in Peru

To achieve its goals, IAPG Peru relies on the collaboration and commitment of its members. It allows generating spaces for participation and joint work over the country. This has allowed the development of articulated plans with civil and student organizations. They have joined the network, which has increased the number of members to approximately 110 nationwide.

IAPG Peru actions include the development of round tables, talks, forums and scientific exhibitions aimed at professionals in geosciences and related branches (Villacorta et al., 2018). The support and cooperation of Peruvian universities such as the National University of San Marcos, National University of Engineering, Antonio Ruiz de Montoya University, as well as geoscientific organizations such as the Geological Mining and Metallurgical Institute (INGEMMET), the Geological Society of Peru and the Peruvian College of Engineers have been fundamental for the development of these activities.

The current moment seems to be particularly favourable to developing actions aimed at improving knowledge of geosciences in Latin America, given UNESCO's interest in promoting research in the region, as well as generating synergies and strengthening integrated actions (Villacorta et al., 2020). Within this framework, distinguished members of IAPG who are also prominent geoscience professionals were invited to participate in international events organized by IAPG Peru since 2014.

3.1 MinerLima: Lima's International Mineral Show

It is a space to disseminate the Peruvian mineralogical heritage, through an exhibition of rocks and minerals, short courses, conferences and workshops (for children and the public), geological excursions and more (Figs. 1 and 2). The most important goal of this event is to educate the Peruvian citizens on geoscience as well as to create links between the geoscience community and society.

Organized annually since 2015 by IAPG Peru, the mineral show involves an expert committee (elected each year) that executes its actions under the IAPG principles. Every year the event captures the interest and has a great reception from the national

Fig. 1 The activities of MinerLima are designed to community in general, including children. Photos Sonia Bermudez (2017)

Fig. 2 Some activities of MinerLima. Photos Sonia Bermudez (2017)

and international public, due to the interesting and extensive programme the exhibition came with. In 2018, about 900 participants took part in the event, more than ever before.

2020 and 2021 editions have been developed in the virtual mode due to the COVID-19 pandemic emergency. The current links to the geoscientific exhibition are:

Website: https://minerlima.wixsite.com/minerlima

Facebook page: https://www.facebook.com/feriainternacionalmineraleslima

Blog: http://feriamineraleslima.blogspot.pe/

3.2 Open Letter on Georisks in Peru, 2017

This initiative responds to the extreme meteorological events occurring in Peru in 2017, addressed to the Peruvian Society with an exhortation to adopt actions. The letter was prepared with the collaboration of IAPG specialists from the International Board.

In this context, to contribute with taking action, the objective of the letter was to provide orientation to stakeholders in urban environmental management and disaster prevention in Peru, and to face georisks. In this calling, it is mentioned the need to take up effective initiatives to promote a professional and scientific discussion about land planning in Peru, encouraging Peruvian society to take actions regarding the geological and climatic conditions that are affecting the country. The letter was very welcomed and published at the international level through the IAPG website. Download the Open Letter here: http://media.wix.com/ugd/5195a5_51e7e71a3a434f80a246a07e4d4da22d.pdf

3.3 Specialized Section in Geoethics, Geological Culture and Society (SEGCGS)

In 2016, the members of IAPG-Peru belonging also to the Peruvian Geological Society (SGP), founded the SGP Specialized Section in Geoethics, Geological Culture and Society (SEGCGS). In 2017, a cooperation agreement between IAPG and SGP was signed. It pursues the collaboration in raising awareness among Peruvian professionals about the ethical performance in the development of their activities as well as promoting joint actions. Nevertheless, even if the SEGCGS would be the body representing IAPG in Peru, due to administrative complications, IAPG-Peru operates as an independent group.

In 2020, a new SGP Board was elected by SGP members, with which IAPG-Peru is discussing to maintain in activity the SEGCGS, allowing to spread more effectively the "Geothical Promise" and to articulate actions within the traditional Peruvian geoscientific community.

3.4 Registration in the Peruvian Public Registers

IAPG-Peru has been legally constituted and registered under Peruvian laws with the Superintendency of Public Registry (SUNARP) on March 4, 2020, as a non-profit organization (NPO) in Peru (certificate of registration number: 14469053).

The advantages of being an NPO in Peru are that IAPG-Peru has now an official legitimacy and credibility to work in the country, to approach different stakeholders in order to develop cooperative actions for continuing to promote the IAPG goals more transparently and effectively. Being an association formally established and recognized by the State, it is possible to interact equally with the other formal state and private institutions. Thus, in this way, the Section can begin the process of forming a greater number of strategic alliances with other relevant institutions to our institutional objectives for the good of the Peruvian geoscientific community and civil society.

The first IAPG-Peru statute was prepared under the IAPG principles. It serves as internal regulations of the Section.

IAPG-Peru is the third IAPG section legally registered in its country, after IAPG-Italy and IAPG-Nigeria.

3.5 Open Letter on Paleontological Heritage

Following the approach of the National Cultural Heritage Law (item Paleontological Heritage: Necessary Details of this book), in 2020, a law proposal on paleontological heritage emerged in which it was initially proposed that paleontological research and fossil administration will be developed by the Peruvian Ministry of Culture. This meant that staff with no experience in palaeontology would be assigned to such work, which is why the proposal was considered detrimental to geoscientific research, as it could generate arbitrariness, bureaucracy and corruption.

To contribute to a solution, IAPG-Peru published an open letter in 2020 expressing its disagreement and calling for action by citizens and authorities to modify the proposal.

Fortunately, the modification of the law proposal took place and on March 4, 2021, during a session of the Peruvian Congress, the Paleontological Heritage Law of Peru was debated and approved, in which INGEMMET has been stated as the body in charge of the administration of paleontological resources. Even though the law is currently in observation, it is considered by the international geoscience community as a success achieved by a collective effort of Peruvian geoscientists (some of whom are members of IAPG Peru and the Peruvian Paleontological Association), congressmen and the Committee on Culture and Cultural Heritage of the Peruvian Congress.

The approval of this law means a change in the study and protection of the paleontological heritage of Peru (Fig. 3).

Fig. 3 Paleontological fossil. Picture credit https://museohn.unmsm.edu.pe/paleover.html

3.6 Creation of the IAPG-Peru Student Group and Relationship with Student Associations

The Student Group has recently been created to promote the research on geoethics and dissemination of Geosciences among young people.

Since student associations have been demonstrated to be efficient collaborators to the network, this group provides a space for Earth science students and promotes its participation in the Section activities. For instance, undergraduate associations such as YES network Peru, Amautas Mineros, Paranoia Geológica and CINGEO have been collaborating on the children's workshops from MinerLima since 2016 (Fig. 4). They were also key actors in the organization of webinars and online courses of MinerLima 2020.

Pre-graduate students are more open to innovative proposals using information and communication technology, since they have grown up using the technological tools needed in today's world. Youth are more concerned about the attention of current issues worldwide such as climate change, global warming and pollution. In this regard, the IAPG-Peru Student Group allows interaction with the young generation, unifying different generations of geoscientists. They develop its activities in the academic, social, political and cultural aspects, seeking to integrate more students who participate and become involved in the projects of the Section.

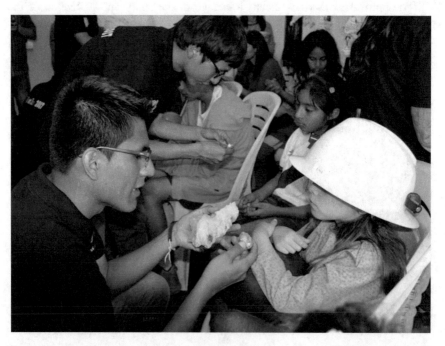

Fig. 4 Member of Student Association "Amautas Mineros" instructing children about minerals during MinerLima 2017. Photos Sonia Bermudez (2017)

3.7 Promotion of Adequate Knowledge of the National Geological Heritage

The promotion of adequate and pertinent knowledge of the national geoheritage is of utmost importance to strengthen the process of sense of belonging and public respect, in addition to the strengthening of the national identity in the citizens. In this sense, through different activities such as seminars and workshops, IAPG-Peru members have been promoting the geological heritage and capturing the interest of the Peruvian geoscientific community. For instance, the workshop organized in 2017 by the Peruvian section and SGP on the Costa Verde hills was an opportunity to visit and explain the geological heritage of the Lima coast. Another opportunity was the webinar on "geoheritage and Earth science education in Latin America and The Caribbean" organized by the Latin American chapter of the International Geoscience Education Organization, in which IAPG-Peru member Jose Cardenas participated showing the experience of the Colca and Andahua valley Peruvian Geopark.

3.8 Partnerships with LAIGEO and IAEG

Due to the IAPG agreement with IGEO, the Peruvian section of IAPG is the Peruvian delegation of IGEO as well. Thus, section members have been actively participating in the international committee of the Latin American chapter of IGEO (LAIGEO). Moreover, IAPG-Peru has been invited to organize the 1st National Geosciences Olympiad in Peru (International Earth Science Olympiad—IESO is the main activity of IGEO). Some steps have been taken, and the organization has already received the official sponsorship of the San Marcos' National University for the organization of the event, and the IAPG Peru board is exploring an alliance with the national university of Piura to implement the project.

On the other hand, the section requested the reactivation of the Peruvian group of the International Association of Geological and Environmental Engineering—IAEG, which had ceased to be active for more than 10 years. The application was approved in March 2021. Since then, a committee of IAPG Peru has joined the group of young IAEG geologist engineers, and a Peruvian Section representative has been invited to participate in the First South American IAEG Conference to be held in Argentina in September 2021. This partnership means more opportunities to continue working alongside international specialists in achieving the goals of the network.

4 Problems Addressed by the Peruvian Section

An outline of the topics related to difficulties in carrying on IAPG-Peru activities is presented in this section, with the purpose to discuss possible solutions.

4.1 Lack of Commitment to IAPG

Some IAPG-Peru board members left their positions, either due to lack of time and interest, as well as conflicting ideas about the network without possibilities to find acceptable solutions for a real benefit of the section. This situation has affected the section actions. Hence, constantly calling new section's members to sustain IAPG Peru's efforts over time is needed. Part of this problem is related to an adequate level of communication with the section's members. Most of them are located in different regions of the country, which creates challenges to efficiently spread the geoethical values and improve the section's visibility at the national level. Action to decentralize sections responsibilities and resources is required to improve these outcomes.

4.2 Members Seeking Profit

It seems that the application of the IAPG no-profit principle is sometimes not fully understood. For example, in previous editions of the MinerLima exhibition, there were problems due to ex-associates' actions, who sought financial gain from the event. The members involved were removed from MinerLima's commission immediately and then from the network since those acts contradicted the geoethical values promoted by the network.

In order to face this problem, IAPG-Peru statute has established that members must not seek profit for themselves or others through the improper use of their positions on committees or commissions. Furthermore, it is necessary to improve mechanisms to avoid this kind of problems, in particular conflict of interests. It requires all members' commitment to defend the principles and values of the Geoethical Promise.

5 Strategies to Strengthen the IAPG-Peru Actions

It is very relevant to continue promoting the interest and basic knowledge about Earth sciences and their importance for the sustainable development of the country. Linked with this goal, the Peruvian section plans to continue its activities and foster increased collaboration among the members of the network and the Peruvian Geoscientist community. Priority will be given to the following topics:

- Working with communities at high risk in hazard-prone areas to promote their knowledge and reflection on the need for prevention.
- Social and environmental responsibility of geoscientists.
- Improvement of geoscience communication.
- Improvement of educational curricula on geosciences.

- Awareness raising on negative research practices: (cases of plagiarism and falsification) and promotion of research integrity and professionalism in the geosciences.
- Promotion of inclusive policies in the geoscientific community, regarding women, minorities and people with disabilities.
- Training of geoscientists on how to share geoscientific information at all levels.

Working with communities can generate the necessary conditions to strengthen responsible citizen participation in extractive activities with environmental and social responsibility. The latter should be understood in conjunction with democracy. Practising democracy in a tolerant and respectable dialogue contributes to the resolution of conflicts and increases levels of governance. It promotes the population to trust in the institutions of the Peruvian State and an efficient negotiation of conciliation.

For instance, the need to improve communication between geoscientists, decision-makers and population is essential due to the conditions of the country. One of the great problems, in this regard, is the historical lack of communication between those stakeholders. This lack of permanent dialogue has generated not only unwell prepared authorities and disinterest in the subject but also a lack of trust and credibility in geoscientists at all levels.

To break this vicious circle, IAPG-Peru will promote the creation of committees formed by specialists from the geoscientific community and local authorities. These committees can help solve problems affecting communities, as soon as the application of geoscience to the management and sustainable development of the country is understood.

Geoscientists can help the population to understand the value of geoscience and teach them about its multiple applications Peppoloni et al. (2012). That is why IAPG-Peru will continue promoting the involvement of geoscientists in teaching activities, popularization of geosciences and its proper application in line with geoethical values of serving society for the public benefit.

> A medium-term task is the development of academic and political strategies in order to support the creation of geoethics courses to train young geoscientists, in the long term to introduce geoethics into the curricula.

References

Bobrowsky, P., Cronin, V.S., Di Capua, G., Kieffer, S.W. & Peppoloni, S. (2018). The emerging field of geoethics. In L.C. Gundersen (Ed.), Scientific integrity and ethics with applications to the geosciences (pp. 175–212) Hoboken, Washington DC: American Geophysical Union, Wiley. https://doi.org/10.1002/9781119067825.ch11

Di Capua, G., Peppoloni, S., & Bobrowsky, P.T. (2017). Cape town declaration on geoethics. *Annals of Geophysics, 60*.

Crespo, B., Míguez-Álvarez, C., Arce, M. E., Cuevas, M., & Míguez, J. L. (2017). The Sustainable development goals: An experience on higher education. *Sustainability, 9*(8), 1353. https://doi.org/10.3390/su9081353

Matteucci, R., Gosso, G., Peppoloni, S., Piacente, S. & Wasowski, J. (2014). The geoethical promise: A proposal, researchGate. *Episodes, 37*(3), 190–191 (2014)

Ministry of Energy and Mines, Peru. (2020). Mining Statistical Bulletin Edition N°05 - p. 08.

Mogk, D. W., Geissman, J. W., & Bruckner, M. Z. (2018). Teaching geoethics across the geoscience curriculum: Why, when, what, how, and where? In L.C. Gundersen (Ed.), *Scientific integrity and ethics with applications to the geosciences* (pp. 231–265) Hoboken, Washington DC: American Geophysical Union, John Wiley and Sons, Inc. https://doi.org/10.1002/9781119067825.ch13

Peppoloni, S., & Di Capua, G. (2012). Geoethics and geological culture: Awareness, responsibility and challenges. *Annals of Geophysics, 55* (3).

Peppoloni, S., & Di Capua, G. (2015). The meaning of geoethics. In: M. Wyss, & S. Peppoloni (Eds.) *Geoethics: Ethical challenges and case studies in earth science.* Elsevier, Waltham, MA. ISBN 978–0127999357, https://doi.org/10.1016/B978-0-12-799935-7.00001-0

Peppoloni, S., & Di Capua, G. (2017). Geoethics: Ethical, social and cultural implications in geosciences. *Annals of Geophysics.*

United Nations (2015). Transforming our world: The 2030 Agenda for Sustainable Development. New York: United Nations

Vasconcelos C., Meléndez, G., Azanza, B., De Miguel, D., Negredo, M. T., Núñez, A., et al. (2018). Enseñanza de las Ciencias de la Tierra, 2018 (26.2) 249–251.

Villacorta Chambi, S. P., Antayhua, Y., Cruz, V., Toledo, C., & Araujo, L. (2018). Geoethics and its impact in Peru, in *Proceedings of the XV Chilean Geological Congress*, 18-21 November, 2018 (2018).

Villacorta, S. P., Sellés, J., Greco, R., Oliveira, A. M., Castillo, A. M., & Regalía, D.A. (2020). LAIGEO, the South American chapter of IGEO and its actions to promote the improvement of Geoscience education in Latin America. *Geological Correlation Series, 35*(2), 67–76.

Background and Current Situation of Geosciences in Peru

Sandra Paula Villacorta Chambi, Carlos Toledo Gutierrez, and Cesar Chacaltana Budiel

1 Introduction

Geoscience can contribute to a better knowledge of the natural processes that shape the Earth's geo-dynamic system and understanding the impact of human action on it. Geoscientists accept that a deep understanding of natural processes requires studying complex dynamics and processes on a global scale. Since the publication of the document "Earth System Science Overview: A Program For Global Change (NASA, 1986)", which examined the Earth as a system, has become the accepted paradigm in geosciences. System thinking has become fundamental in developing Earth literacy (Locke et al., 2012).

In this framework, the participation of the geoscientific community in Peru is crucial to help in the management of socio-environmental problems linked to mining activities, land planning and corruption.

Since corruption plays a negative role in the formulation of geological research in developing and underdeveloped countries (Desikachari, 2015), it is essential to combat it using a geoethical approach by geoscientists (Lambert, 2012). In achieving

Electronic supplementary material The online version of this chapter (https://doi.org/10.1007/978-3-030-86731-7_2) contains supplementary material, which is available to authorized users.

S. P. Villacorta Chambi · C. Toledo Gutierrez (✉)
Charles Darwin University, Casuarina, Australia
e-mail: carlos.toledo@uarm.pe

S. P. Villacorta Chambi
e-mail: sandra.villacorta-chambi@cdu.edu.au

C. Chacaltana Budiel
Antonio Ruiz de Montoya University, Pueblo Libre, Peru
e-mail: cchacaltana@ingemmet.gob.pe

Instituto Geológico Minero Y Metalúrgico, Lima, Peru

this goal, the social implications of geoscientific research must be considered, which contributes to improving the impact and quality of these research programs. In this regard, a wide discussion between geoscientists and citizens would also be promoted to promote an ethical approach to the problems that affect these nations.

2   The Peruvian Geoscience Community

The Peruvian geoscientific community includes public and private entities. Among the public ones, the Geographical Society of Lima, the Geological Mining and Metallurgical Institute (INGEMMET), the Geophysical Institute of Peru, the geological chapter of the Engineering College and Peruvian universities with faculties of geology and geography stand out in 11 regions of Peru. The Peruvian National Council for Science, Technology and Innovation is also involved here. Private entities include the Geological Society of Peru, the Association of Peruvian Geologists and the Institute of Mining Engineers. To have an overview of these institutions, it is needed to know how they were created.

2.1   *History of Public and Private Geoscience Institutions*

Historically in Peru, research activities were determined by the political situation. Due to the confrontation between power groups that administered the national economic system. In this period, geoscience professionals prioritized personal gain over promoting national identity research. Then there was a vigorous campaign to make Peru's mineral wealth known to the world, which led to the creation of institutions controlled by the academics of that time.

In 1825, Don Mariano Eduardo De Rivero y Ustáriz was appointed general director of Mining and was commissioned to found a School of Mines in Lima and to work on the planning and creation of the country's first "National Museum of Natural History, Antiquities and History" (Deustua, 2017).

With a growing demand for engineers, and due to the rapid development of mining and communications in Peru under the José Rufino Echenique government, French engineers were hired in 1852 to outline a Central School of Civil Engineers (named this way to distinguish itself from the military one). This task was finished in 1857 with regulations of the "State Corps of Engineers and Architects". Where mining engineers among other functions gathered data and collected geological and mineralogical samples for the future creation of the Peruvian geological map. Its regulations were approved in 1872 (López, 2012., p. XVII-XXII), and the "Mining and Manufacturing Section" of that institution was created to continue making a geological map and inventorying mineral resources in Peru (Tamayo et al., 2017).

In 1883, after the defeat against Chile in the "nitrate war", Peru, indebted in millions of pounds sterling with England, gave a greater impetus to the exploration of mineral resources for its economic reconstruction. In this sense, the Governing Board of 1886 organized an advisory group of experts formed by the Civil Engineers School's first graduating class to develop among other activities, "the creation of collections of minerals, rocks and fossils" (Corps of Mining Engineers of Peru, 1902). In this scenario, the intellectuals and academics, under the philosophical current of positivism, had a discourse where science, education, European support and capital were fundamental for the reconstruction of the State. Along with that trend, some intellectuals of the time embraced the oligarchy, represented by the Civil Party. That meant a "relative political stability, economic modernization and consolidation of the civilian elite" (Cueto, 1989) called by the geographer Modesto Basadre, the "Aristocratic Republic". Some institutional leaders such as Luis Carranza were part of that movement. In 1895, Carranza was a member of the Civil Party board and also of the Governing Council of the Executive Power (López-Ocón, 2001). He gave financial support to the geological research (Caravedo, 1941), for example, José Balta's remarkable geological investigation, which acknowledged and designated the fossil Cruziana *Carranzae* in his honor in 1897.

In this framework, some institutions arose. First, the Geographical Society of Lima (SGL) was founded in 1891 with Luis Carranza as its first board president. He designed a broad research science program that developed specialized groups such as the "Technical Commission of Mineralogy, Geology and Paleontology", under the presidency of Leonardo Pflücker y Rico. One of the commission's aims was to publicize Peruvian mineralogical wealth, its geological characteristics and the study of fossils (Fig. 1).

In 1896, the Ministry of Public Works was created, and in 1902, the Corps of Mining Engineers, under the government of José Lopez de Romaña y Alvizuri. Its Main Office carried out the analysis and classifications of minerals, rocks and fossils and had libraries and museums of Mineralogy, Geology and Paleontology. Their staff was constituted of specialists in Mineral Chemistry, Paleontology and Petrography. Jose Balta was its Development Director and called José Bravo as the Laboratory Cabinet Head (Corps of Mining Engineers of Peru, 1902). According to Cueto (1989), Balta was "a notorious Freemason member of the Grand Lodge of Peru" and represented the direct relationship between the Academy and the Civil Party. This relationship explains the financial capacity of geoscientists belonging to the civilian elite, whose members had all the facilities to access the results of European research that were later replicated in Peru.

On the other hand, in 1911, Antonio Raimondi's fossil collection had the SGL sponsorship for its inclusion in Volume VI of the series entitled "Peruvian pale-ontology" under Carlos Ismael Lisson Beingolea supervision. Since 1911, Lissón generated systematic records of the fossil fauna and flora that allowed him to make the first paleontological map of Peru (López-Ocón, 2001). Establishing links between his work and the elite of the time, Lisson named one of the fossils discovered *Perisphinctes pardi* in 1906 in honor of the Peruvian President José Pardo (Cueto, 1989). Later, in 1922, the Department of Terrestrial Magnetism of the Carnegie Insti-tution for Science of Washington founded the Huancayo Geomagnetic Observatory in the Junín region (IGP, 2019). At the same time, with the support of Jorge Broggi

Fig. 1 Conference Room of the Geographic Society of Lima in 1895, during the Presidency of Luis Carranza. Photo taken from Bulletin IV, Geographical Society of Peru (1895)

and Aurelio Masías, Lisson founded the Geological Society of Peru (SGP) at the SGL facilities (Geological Society of Peru, 1925). In 1944, SGP incorporated Dr Manuel Prado as honorary President (head of the Peruvian state at that time), who announced his intention to create a technical institution for the elaboration of the Geological Map of Peru.

To promote mining, the "Peruvian Institute of Mining Engineers" was created in 1943, and given the gradual need to increase the scope of the State in mining exploration, in 1944, a department of the Corps of Mining Engineers was transformed into the "Peruvian Geological Institute". Under the supervision of Jorge Broggi as head of that entity, the geological investigation continued with the support of international scientific cooperation. As an example of this, the Carnegie Institute for Science managed the Huancayo Geomagnetic Observatory, (which changed its name to the Huancayo Geophysical Institute (IGH) in 1947), until 1962 when it became an autonomous body of the Peruvian State (IGP, 2019).

When the Peruvian government decided in 1945 to cede a building shared by the Peruvian Geological Institute and the Geological Society of Peru, the strategic association of the State with private geoscientific associations became evident.

Fig. 2 Salvador Mendívil Echevarría, promoter of the Geology and Mining Service and supervisor for the Geological Map at a 1: 1,000,000 scale. The First Latin American Congress of Geology took place in 1970. This is the "Golden Age" of the Peruvian Geology. Photos Cesar Chacaltana (1970)

In 1950, the Corps of Mining Engineers and the Geological Institute were reintegrated under the name of "National Institute for Mining Research and Development" (INIFM). This new institution created the departments of "Geological Research" and "Geological Maps".

With this background, in 1960 during the government of President Manuel Prado, the National Geological Chart Commission (CCGN) was created, under the supervision of Dr Isaac Tafur Hernández, who established the beginning of systematic geological mapping at scale 1: 100,000. In 1963, Salvador Mendívil (Fig. 2) proposed the creation of the "Mining and Geological Survey" (SGM), which was possible after merging CCGN and INIFM in 1965. Once the SGM was established, the Department of Paleontology was created by Dr Alfredo Pardo Argüedas, who organized the paleontological information in order to establish the first stratigraphic unit map of Peru. SGM's first achievement was the elaboration of the Geological Map of Peru at a scale of 1: 1,000,000, with its geological time units based on the paleontological information collected, which represented the first Paleontological Database of Peru (Fig. 3).

On the other hand, in 1962 to replace the IGH, the Geophysical Institute of Peru (IGP) was created. For this, the IGH executive headquarters moved from Huancayo to Lima. It is worthy to note that in 1964, the IGP sent 12 Peruvian engineers to foreign universities to obtain a PhD degree; a unique effort within the Peruvian public administration at the time (IGP, 2019).

In 1968, the Energy and Mines Ministry and the National Research Council (CONI) were created. The former began its activities in 1969 with the electricity, hydrocarbons and mining subsectors, while the latter with the mission to promote, coordinate and guide scientific and technological research in Peru (CONCYTEC, 2016).

Fig. 3 Alfredo Pardo Argüedas, founder and head of the Department of Paleontology of the National Geological Chart Commission. The lower image shows the Paleontological Record Book, which was the first Paleontological Database of Peru in 1965. Photos Cesar Chacaltana (n.d)

It is also worth mentioning that given the influence of the SGM at the continental level and the links generated while making the Peruvian geological map, Peru hosted the First Latin American Geology Congress in 1970. The coordination was carried out by the Energy and Mines Ministry with support from the Association of Geologists in Peru, the Geological Society of Peru and the Peruvian Institute of Mining Engineers. This event showed the high level of mutual support between state and private geoscientific institutions.

Then, in 1973, the "Scientific and Technological Institute in Mining" (INCITEMI) was created, whose function was to promote, coordinate, guide and carry out research and scientific/technological support to the mining industry. In 1975, the Geology and Mining Service was renamed as the "Institute of Geology and Mining" (INGEOMIN) and in 1979, INCITEMI and INGEOMIN were merged under the name "Geological Mining and Metallurgical Institute" (INGEMMET) and it became part of the Energy and Mines Ministry of Peru.

In 1990, under Alan García's first presidency, Peru suffered the worst economic crisis in its history as a Republic, with high repercussions at the social level. This period is characterized by unemployment and the uncertainty of working without a permanent contract. This period was mired by corruption that affected professional circles, and the geoscience community was not exempt. Faced with this situation, a series of measures were taken in Peru to encourage investment in the mining sector (Galarza, 2004). This had the effect of the so-called "geological exploration mining boom", which arose in Peru. In that period, the professional schools of Geological Engineering were suddenly overwhelmed with new students, as there was a labor market in those industries. The institutions related to mining activity such as INGEMMET were modernized. The institution focused on completing the National Geological Chart at 1: 100,000 scale, the elaboration of 137 bulletins and the generation of a database of approximately 15,000 fossils (Fig. 4).

In 2005, after the economic crisis, INGEMMET started a regional geological research program that included projects in different professional branches of the geosciences, such as Geo-risks and Hydrogeology. Under the supervision of General Director Jose Machare, the strategic management of geoscientific information started with a view to comprehensively modernize the institution. For this, systematic field and office work began with the financial support of the Energy and Mines Ministry, which would ensure availability of the country's geoscientific information, always focused on promoting private investment (Lay, 2005). In 2007, INGEMMET was merged with the "National Institute of Concessions and Mining Cadastre" (INAAC), retaining the name INGEMMET. Since 2008, under scientific programs linked to the national plan of research, this institution has strengthened its team of specialists, incorporating new young researchers. Today, this mining sector entity has a significant budget for its research activities. However, these studies continue to prioritize the evaluation of mineral resources and related issues.

On the other hand, between 1985 and 1991, the Ministry of the Presidency and the Ministry of Education fought to incorporate CONI and changed its name to the Peruvian National Council for Science, Technology and Innovation (CONCYTEC). Between 2005 and 2010, its functions were reviewed and restructured. In 2012, it was assigned to the Council of Ministers Presidency and reorganized to strengthen its capacity as the governing organization for the development of Science, Technology and Innovation in Peru (CONCYTEC, 2016). At present, this state body is in charge of directing, promoting, coordinating, supervising and evaluating the actions of the Peruvian State in the field of science, technology and technological innovation. It also guides and executes actions to promote the scientific and technological development of the country. Through an annual strategic plan, this entity prioritizes the granting of funds, bonuses to researchers, postgraduate scholarships, amongst others (Machare, 2018; Unique digital platform of the Peruvian state, 2020).

On the other hand, IGP is now a decentralized public body under the Ministry of the Environment of Peru, which is responsible for the detection of potentially destructive natural phenomena (earthquakes, tsunamis, volcanic eruptions, debris-flows, amongst others). This organization collaborates with Peruvian university's

Fig. 4 The upper photograph is a view of the current building front of INGEMMET. The lower photograph is part of its Paleontological Collection, which has supported the elaboration of the National Geological Chart. Photos Cesar Chacaltana (2020)

geoscientific research programs and supports undergraduate, masters and PhD thesis of young bachelors in geosciences (IGP, 2019).

2.2   The Universities

In 1876, the School of Civil Constructions and Mining Engineers was founded by the Polish engineer Edward Jan Habich. In 1891, its structure included mineralogical, geological and paleontological museums. Outstanding professors at the time were Teodorico Olaechea, Sebastián Barranca, Carlos Lisson, Eugenio Weckwarth and José Bravo being the last professor in charge of the mineralogy laboratory and the paleontology museum (López, 2012). In 1955, the School changed its name to the National University of Engineering and its departments according to academic specification became faculties. In the 1960s, the geology department was created, which was in common with the faculty of petroleum engineering and the mining one (López Soria, 2003).

In 1967, the Faculty of Geological, Mining and Metallurgical Engineering was created, and Rosalvina Rivera was assigned to Paleontology, Micropaleontology, Stratigraphy, Historical Geology and the Geology of Peru. She was in charge of the courses until her retirement in 1974. Rivera updated the organization of the UNI Paleontological Museum and contributed along with the German geologist Georg Petersen to create the UNI's Mineralogical Museum (Alleman & Rivera-Charun, 2011).

At present, it has a renovated Museum of Mineralogy and Paleontology thanks to the support of Guido del Castillo, who was a benefactor and chairman of the UNI Board of Trustees. The Museum has a valuable collection of 6000 samples, of which 1000 samples have been classified. It also has a valuable paleontological collection, amongst which are unique specimens of fossil mammals (Fig. 5).

On the other hand, the San Marcos's National University's (UNMSM) Faculty of Sciences was created in 1850 and then in 1866 its Dean, Antonio Raimondi, organized courses in Mineralogy, Geology, Botany and Zoology in the Physical Sciences Section of the Faculty of Sciences. A year later Sebastián Barranca Lovera took charge of the subjects of Mineralogy, Geology and Paleontology for the following 25 years (Saco, 1995). Since the early 1900s, the National University of San Marcos' (UNMSM) geological studies were developed under the supervision of Barranca Lovera, who was also the head of the respective labs and cabinets (Ribeyro, 1876, Chacaltana, 2018). Between 1918 and 1932, Carlos Lisson directed the subjects of Geology and Paleontology at UNMSM and later in 1935 founded the academic specialty in Geology as part of the Physico-Geological Sub-Section of the Faculty of Sciences. This section became the first Peruvian School of Geology (Lisson, 2008; Saco, 1995). An important institution from this university is the "Javier Prado" National History Museum, which includes Mineralogy and Petrology sections as well as Invertebrate Paleontology and Paleobotany. It includes the Antonio Raimondi, Bernardo Boit and Axel Olsson collections, which, in total, add up to almost 4,000

Fig. 5 Front of the Faculty of Geological Engineering of UNI. The photographs at the bottom show an exhibition room of the UNI's Museum of Mineralogy and Paleontology (left) and a large fossil vertebrate (right). Photos DePeru.com. (n.d)

samples with 20,000 specimens. In addition, the Vertebrate Paleontology section's scientific collection has become a continental reference today (Fig. 6). During its almost 20 years of existence, it has managed to arrange studies and publish on at least 65 new species (Benites-Palomino et al., 2018).

In recent times, the geoscienctific community has grown due to the inclusion of bachelor's degrees in geology and geophysics in several public and private universities in other Peruvian regions since the 1980s (Chap. 7 of this book). Young geoscientists in training tend to create student organizations (see the heading "Relations with student associations" in Item 1.4.7) and are more closely related to new technologies and international trends in research.

2.3   International Cooperation

As this chapter has shown, international cooperation since the Republic era has made a strong contribution to geological research in Peru. Foreign Entities that usually cooperate with Peruvian geoscientists are: National Council for Scientific and Technical Research (CONICET) of Argentina, La Plata National University

Fig. 6 View of the UNMSM National History Museum facilities. Below: fossil's vertebrate showroom. Photos Oviedo (2021)

(Argentina); Geológical and Mining Institute of Spain; Higher Council for Scientific Research of Spain; American Museum of Natural History, New York (USA); Florida University; Smithsonian Tropical Research Institute; Loma Linda University (California); Washington University; Natural History Museum of Los Angeles County, (California); Università di Pisa, Italy; Museum National d'Histoire Naturelle, Paris, France; Institut de Recherche pour le Développement (IRD), Lima; Université de Montpellier, France; Université Paul-Sabatier, Toulouse; French Institute of Andean Studies; Institut Royal des Sciences Naturelles de Belgique, Bruxelles and the Natuurhistorich Museum, Rotterdam. The Peruvian research cooperation is provided through agreements and/or memoranda of understanding with public and private universities or with research entities such as IMGEMMET and IGP.

3   Main Career Options to Peruvian Geoscientists

Nowadays, the majority of the Peruvian geological engineers (92%) opt for the exploration of mining and energy resources as well as consulting services at companies in environmental geology and geological risk management. Only the remaining 8% are dedicated to teaching and research (Machare, 2018). More recently, the branches Peruvian geoscientists are looking at are hydrogeology, remote sensing and GIS.

Those geoscience experts who have an academic career in Peruvian universities do not have an easy pathway forward due to the current bureaucracy, organizational deficiencies, disarticulation between deans and schools, as well as low budgets and salaries. Due to the low remuneration, many professors dedicate themselves to working as private consultants as well. This affects their ability to dedicate solely on teaching activities.

4   Main Issues

Being linked to the mining industry since the Incan Empire and from the last decades, mining activity has transformed Peru into one of the largest producers of metallic minerals in the world (Garcia, 2011). At the moment, the country enjoys a good worldwide position as a destination for mining and exploration due to the prolific capital expenditure related to mineral resources (APCSG, 2011). Mining was the largest contributor to GDP in 2013 at 13 % of Peruvian production. However, Peruvian mining is expected to lose its economic power in the coming years due to a decrease in prices, reduced Chinese demand for mineral resources, and slowing foreign investment due to global monetary policy (Lipinsky et al., 2015). However, Peru's large mining industry is a concern for a country with a fast-growing economy associated with extractive enterprises. In this sense, continuing uncontrolled mineral exploration without a geo-ethical framework remains a problem in some Peruvian regions (Gardner, 2012; Langeland, 2015). It has created social conflict involving farming communities, which continue to protest against the contamination of fluvial systems, demanding its protection (Li, 2013, 2016). This topic will be covered more extensively in Chap. 4.

Geosciences in Peru involve more geoscientists in geohazard studies than in the past due to the particular geo-dynamic configuration of the country (Villacorta et al., 2012). For instance, during the summer of 2017, Peru was affected by extreme meteorological events, which caused a lot of damage and even deaths (Rodríguez-Morata et al., 2019). However, the biggest issues so far derive from the urban improvisation, informality and mismanagement of municipal authorities (Bankoff, 2017; Macedo et al., 2014; Practical actions, 2017). There are political and social processes such as land planning prior consultation and social licensing, among other subjects, that demand the active involvement of scientific and technical institutions including geoscientists. This issue is explored further in Chap. 3.

On the other hand, according to Macedo et al. (2014), the most important problems affecting the growth of geosciences in Peru are the following:

- The lack of geoscientific communication with the population.
- The lack of knowledge and interest in topics related to Geosciences and scientific information.
- The continuous changes in government authorities, which causes delays in developing long-term policies.
- The lack of collaboration between geoscientists, journalist and communicators.

Regarding the main problem in the education of geosciences, it is worth highlighting that over the last 20 years, geosciences have not been included as part of the standard school curriculum. This issue will be discussed in Chap. 7.

Other geo-ethical problems in Peru include a variety of aspects. First, Peruvian research policy and evaluation processes are inadequate and the access to research funding is limited (Cáceres and Mendoza, 2009). For example, since 2016, CONCYTEC's evaluation system for funding research has become far more demanding and requires additional components for applications (Machare, 2018). The national budget for research support has been slightly increased, but it is not enough to promote the desired culture of scientific research in Peru (Bermudez García, 2014; Mezarina & Cueva, 2017). The greatest concern is that Earth sciences are generally not considered a priority for research by CONCYTEC. For example, in 2018, geoscience topics were distributed as subtopics of other areas within the natural sciences (Machare, 2018).

Furthermore, there are several cases of malpractice in research, such as plagiarism. For example, correct referencing is not a regular practice in research activities in certain specialities such as the geosciences. The authors of prior research investigations are not quoted in reports of public entities (Sucapuca, 2017). Moreover, funding is requested to investigate what research had already been completed in the past by other Peruvian institutions instead of developing new studies. Finallly, the activity of pseudo-scientists has been accepted by a certain sector of the Peruvian scientific community. Some self-proclaimed geoscientists also received research funding and were involved in scandals related to the organization of international congresses (see Chap. 5).

5 Discussion

Paleontological studies have shown that life is a function of the Earth System and that the human consciousness is the result of an evolutionary process of biological development. In general, social organization is the source of moral values, which affect ethical conduct in science. In the fields of scientific practice, codes of conduct have always moved from moral to ethical, that is, from action to reflection. Countries like Peru, under the current capitalist economic system (which has colonial roots), are dependent on foreign technologies and many geo-ethical contradictions have arisen.

As this chapter has attempted to show, despite their being public institutions dedicated to geoscience research in Peru, these organizations belong to ministries with varying competencies in the field. Also, worth noting is that those organizations that initially worked together now carry out separate research programs and activities. There is no national strategic plan to promote new research approaches in geoethics, which for the most part is unknown to the Peruvian general population. Moreover, it is neither a topic found in the study plan of the Earth science's schools nor professional faculties in the country. This is evidence that this subject has not yet been embraced by Peruvian institutions, but could be a new area of research for contemporary geosciences in Peru. The obvious lack of philosophical education in geosciences limits the level of awareness of the importance of geoethics in the professional practice of geosciences. Geoscience education is stuck in the gap between the curricula of the country's universities and worldwide trends.

The geoethical approach should guide ethical work in geoscience research (Di Capua et al., 2016; Peppoloni, & Di Capua, 2012) and could help to promote the socio-political change required for sustainable development in Peru (Piscoya, 2016; Villacorta et al., 2016). This is why the participation of geoscientists in the country's political activity should be encouraged. Their role is particularly relevant in promoting responsible and ethical behavior in professional practice based on geoethical values. Practicing geoethics will also assist Peruvian citizens to reduce their distrust of geoscientists who are incorrectly believed to be linked to the elite and mining companies. This will also contribute to the sustainable future development of countries like Peru.

Although the Peruvian Geoscientist community used to belong to a powerful elite for a long time, today, this is very far from reality as this community is now primarily integrated with young professionals from all Peruvian regions. Knowing that geoscientists of the twenty-first-century practice geoethical values in the pursuit of the Earth's sustainable development will help citizens to value Geosciences, the work of Earth science specialists, and to get involved in creating change through political process.

In this context, some new processes are required. First, the identification of stakeholders involved in the implementation and promotion of benefits from applying a geoethical approach. If the development of policies is left to uninformed stakeholders, the vicious cycle of underdevelopment in countries like Peru will continue. Instead, if they are familiarized with better sustainable development strategies, better decisions will be made in those countries to produce a global paradigm shift. This is also linked with generating a political promotion to demonstrate the relevance of geoethics in the Peruvian educational curricula and to ensure that the next Earth sciences generations will incorporate the geoethical approach in their professional activities.

Second, high-quality scientific communication developed in coordination with government institutions is needed. To achieve this, the use of adequate channels to spread Geosciences efficiently will help it to gain visibility in society. It is also needed to modernize the statutes of scientific organizations, thinking about opening up society and improving how the Peruvian Academia is involved in the processes of those institutions.

Third, the training on the geo-ethical approach should be considered. Whether as part of the curricula or not, this will ensure that geoscientists become active agents of change. In addition, the practice of the geoethical promise can strengthen Peruvian geoscientists, key competencies for adapting to the modern world. These are mainly interpersonal skills that will help them not only to master their branch but also to become proactive in successfully solving issues like social-environmental problems. It will also facilitate their communication with authorities, increase their participation in domestic political activity and allow participation in committees making decisions for the benefit of Peruvian society.

Finally, the actions of the International Association for the Promotion of Geoethics—IAPG, through its Peruvian section, should continue enabling citizens to understand and practice geoethics in Peru. Their intervention will make it possible to democratize Geosciences in Peru. The usefulness of bringing science and society closer together has been demonstrated in Community Relations Management. Today, this is crucial to the success of projects, especially in the mining sector.

6 Conclusions

On balance, the Peruvian Geosciences Academy was historically linked to the elite, which supported the decision-making political process. In the beginning, the purpose of the academy was to obtain geological knowledge about the Peruvian territory and its resources. In recent times, the geosciences' community has been diversified with professionals from around the country.

Despite the recent introduction of geoethics, Peruvian geoscientists still face a public bias today, and the geoscientific institutions demonstrate disjointed work. Proposals to improve the visibility of Geosciences in Peru, such as modernizing these organizations, promoting the participation of geoscientists in the political process of the country, and preparing the twenty-first century geoscientists with a geoethical framework. All these actions will allow efficient handling of the complex problems affecting the science–society interface in Peru. This appears to be the only possible way to successfully contribute to the dissemination of geoscientific culture within our society.

References

Alleman, V., & Rivera-Charun, M. (2011). A brief review of Dr. Rosalvina Rivera. *Bulletin of the Ricardo Palma University's Natural History Museum, 12*(17). Retrieved February 24, 2021, from https://www.urp.edu.pe/pdf/id/3263/n/boletin-museo-2011.pdf.

APCSG (2011). *Office for the Prevention of Social Conflicts and Governance: Reports*. Retrieved February5, 2011, from http://www.defensoria.gob.pe/conflictos-sociales-reportes.php.

Bankoff, G. (2017). Living with hazard: Disaster subcultures, disaster cultures and risk-mitigating strategies. In *Historical disaster experiences* (pp. 45–59). Springer International Publishing.

Benites-Palomino, A., Bellido-Valverde, D., Olmedo-Romaña, G., Burga-Castillo, M., Soria-Hilares, C., & Aliaga-Castillo, A. (2018). The department of vertebrate paleontology of the museum of natural history UNMSM. What have we learned in 20 years? In *II International Symposium on Paleontology of Peru. Book of Abstracts* (pp.176–178).

Bermúdez García, J. E. (2014). Scientific research in Peru: A critical success factor for the country's development.

Cáceres, C. F., & Mendoza, W. (2009). Globalized research and "national science": The case of Peru. *American Journal of Public Health, 99*(10), 1792–1798.

Caravedo, B. (1941). Luis Carranza (Biographical essay). Lima, Printing house of the Hospital Victor Larco Herrera.

CONCYTEC (2016). *Institutional memory 2016*. Retrieved February 20, 2021, from http://portal.concytec.gob.pe/images/publicaciones/memoria_institucional_2016.pdf.

Corps of Mining Engineers of Peru. (1902). Official Documents, Bulletin N ° 1, 47 pp.

Cueto, M. (1989). Scientific excellence in the periphery. Scientific activities and biomedical research in Peru 1890–1950. GRADE-CONCYTEC, 230 pp.

Deustua J. (2017). Society, science and technology: Mariano De Rivero, mining and the Birth of Peru as a Republic, 1820–1850. http://www.scielo.org.pe/pdf/apuntes/v44n80/a02v44n80.pdf.

DePeru.com. (n.d). *UNI Museum of Mineralogy and Paleontology* [Photograph]. Retrieved from https://www.deperu.com/cultural/museos/museo-de-mineralogia-y-paleontologia-de-la-uni-4382.

Di Capua, G., Peppoloni, S., & Bobrowsky, P. (2016). The cape town statement on geoethics. In N. Bilham, M. Bohle, A. Clay, E.H. Lopera, D. Mogk (Eds.), *IAPG - international association for promoting geoethics*. http://www.geoethics.org/ctsg.

Desikachari, V. (2015, April). The relevance of geoethics to under-developed and developing Nations with special reference to India. I. In *EGU general assembly conference abstracts* (Vol. 17).

Galarza, E. (2004). *The economics of natural resources*. Pacífico University. Research Center.

Garcia, M. J. (2011). *The role of employee capacity building in reducing mining company-community conflicts in Peru* (Doctoral dissertation, University of British Columbia).

Gardner, E. (2012) Peru battles the golden curse of Madre de Dios. *Nature, 486*, 306–307; Geological Society of Peru (1925). Minutes of the SGP Sessions. Session I (p. 5–7, Vol. 1), 126 pp.

IGP (2019). *Management transfer report*. Lima, Peru. Retrieved from https://www.minam.gob.pe/transparencia-/wp-content/uploads/sites/48/2019/11/INFORME-TRANSFERENCIA-DE-GES TION-IGP.pdf

Lay, V. (2005). 2005 annual report. Mining and metallurgical geological institute. 70.

Lambert, I. B. (2012). Geoethics: A perspective from Australia. *Annals of Geophysics, 55*(3).

Langeland, A. L. (2015). *Impact of alluvial artisanal and small-scale gold mining in the Madre de Dios River Basin, Peru: total mercury levels in human and farmed fish populations*. Doctoral dissertation, University of Michigan.

Li, F. (2013). Relating divergent worlds: Mines, aquifers and sacred mountains in Peru. *Anthropologica* (pp. 399–411).

Li, F. (2016). In defense of water: Modern mining, grassroots movements, and corporate strategies in Peru. *The Journal of Latin American and Caribbean Anthropology, 21*(1), 109–129. ISSN 1935-4932, ISSN 1935–4940.

Lipinsky, F., Ross, K., Tashu, M., Vtyurina, S., & Fenochietto, R. (2015.) "Peru: Selected issues." International Monetary Fund. IMF Country Report No. 15/134, May 2015.

Lisson, C. (2008). Lisson, the wise man who measured time. In: Semblanzas de una estirpe Limeña. The Lissons (pp. 33–60). Caracas. https://fliphtml5.com/qdbob/moqm.

Locke, S., Libarkin, J., & Chang, C. Y. (2012). Geoscience education and global development. *Journal of Geoscience Education, 60*(3), 199–200.

López, J. (2012). History of UNI, (Vol. I). The founding years (1876–1909). UNI University Press, 337 pp.

López-Ocón, L. (2001). The geographical society of Lima and the formation of a national science in Republican Peru. Terra Brasilis (Nova Série) Revista da Rede Brasileira de História da Geografia e Geografía Histórica. 22 pp.

López Soria, J.(2003). Brief history of UNI. Lima: National University of Engineering. Edited By UNIPETRO ABC SAC. Retrieved February 24, 2021, from http://www.fondoeditorial.uni.edu. pe/Resumen%20de%20la%20Historia%20UNI.pdf.

Macedo, L., Villacorta, S., Vasquez, S., Mariño, J., & Di Capua, G. (2014). Geoscientific communication problem with communities for disaster prevention and land planning in Peru. In *Engineering geology for society and territory* (Vol. 7, pp. 81–83). Cham: Springer.

Machare (2018). *Organization of geoscience research in Peru and worldwide* [video]. Lecture at "Geocientíst Friday" organized by INGEMMET, 18–05–2018. Retrieved from https://www.you tube.com/watch?v=r3DauJhWsN4.

Marco, G., Carlos, J., Sá, A. A., García Bellido Capdevila, D., Budiel, C., & Augusto, C. (2017). Recent geoethical issues in Moroccan and Peruvian Paleontology. *Annals of Geophysics, 60*, Fast Track 7.

Mezarina, J., & Cueva-UPCH, S. (2017). In Peru, science is advancing, are its scientists advancing? *Economía y Sociedad, 91*, 26.

NASA Advisory Council. Earth System Sciences Committee. (1986). *Earth system science overview: a program for global change.* National Aeronautics and Space Administration.

Oviedo, O. (webmaster) (2021). *UNMSM Natural History Museum* [Photograph] Retrieved from https://www.museosdelima.com/museo-de-historia-natural-unmsm/.

Peppoloni, S., & Di Capua, G. (2012). Geoethics and geological culture: Awareness, responsibility and challenges. *Annals of Geophysics, 55*(3).

Piscoya, J.C. (2016). Geoethics: Importance of democratizing Geosciences in Peru. Lecture in "Miércoles Geológicos". Peruvian Geological Society. Retrieved April 3, 2016.

Practical Actions. (2017). PERU 2017 Risks, Disasters and Reconstruction.

Rodríguez-Morata, C., Díaz, H. F., Ballesteros-Canovas, J. A., Rohrer, M., & Stoffel, M. (2019). The anomalous 2017 coastal El Niño event in Peru. *Climate Dynamics, 52*(9–10), 5605–5622.

Saco, O. (1995). Historical review of the specialty of Geology commemorating 60 years of its creation at UNMSM. Retrieved from https://fondoeditorial.unmsm.edu.pe/index.php/fondoedit orial/catalog/book/96.

Sucapuca. (2017). Activities of the IAPG-Peru: Promoting an adequate application of Geosciences. Lecture in "Miércoles Geológicos of SGP". Retrieved Augest 23, 2017.

Tamayo, J., Salvador, J., Vásquez, A., & Zurita, V. (2017). The mining Industry in Peru: 20 years of contribution to the growth and economic development of the country. Lima, Peru. Osinergmin, 316 pp.

Villacorta, S. P., Fídel, L., & Zavala, B. (2012). Susceptibility mass movements map in Perú. *Journal of the Argentine Geological Association, 69*(3), 2012.

Villacorta, S .P., Perez-Puig, C., Piscoya, J., Villacorta, G., Di Capua, G. (2016). Promoting Geoethics in Peru: The activities of the IAPG Peruvian Section. In: *Conference: 35th International Geological Congress.* https://doi.org/10.13140/RG.2.2.10911.15521.

Unique digital platform of the Peruvian state (2020). *National Council of Science, Technology and Technological Innovation. Presidency of the Council of Ministers.* Retrieved from https://www. gob.pe/4141-consejo-nacional-de-ciencia-tecnologia-e-innovacion-tecnologica-que-hacemos.

Georisks and Their Implications for the Peruvian Society

Pedro Isique Chaname and Sandra Paula Villacorta Chambi

1 Introduction

Due to the complex geological and climatic characteristics and being included in the so-called Ring of Fire, Peru is constantly affected by the effects of geodynamics processes, which increase the risk of disasters. However, Peruvian citizens have adapted to the territory conditions for more than 5,000 years. Caral (Fig. 1) and Moche Culture (150 AD–800 AD), located in the coastal area of the country, are examples of that adaptation, which was achieved during the Inca Empire (1200–1533 of our era), a high level of human development.

This is demonstrated by the establishment of safe and resilient human settlements, almost exclusively in the Andes (1533 to the present). There was mainly urban development on the western coast of Peru. Since then, cities have been built in semi-arid coastal valleys (today 58% of Peru's total population lives there, according to the INEI Census, 2017). Due to water scarcity, facilities were built to transfer drinking water from the upper parts of the Andes Mountain Range to the coast.

Nevertheless, it is considered that the risk of disasters increases in Peru as the population rises (nowadays, more than 32 million inhabitants according to INEI, 2017). This problem is more evident since the second half of the twentieth century and involves a disorderly growth due to the migratory process from the countryside to the different coastal urban cities, such as Lima (Villacorta et al., 2015). In recent

Electronic supplementary material The online version of this chapter (https://doi.org/10.1007/978-3-030-86731-7_3) contains supplementary material, which is available to authorized users.

P. Isique Chaname (✉)
PIASA Consultores SA, Lima, Perú
e-mail: pisique@piasaconsultores.com

S. P. Villacorta Chambi
Charles Darwin University, Casuarina, Australia
e-mail: sandra.villacorta-chambi@cdu.edu.au

Fig. 1 Caral city. Takenfrom: viajes del perú.com (2017)

decades, this migration has dramatically increased (between 1980 and 2000 due to the terrorism in rural areas) and endangered the loss of ancestral indigenous knowledge about the adaptation to the natural conditions of The Andes.

Hence, the importance of developing geoscientific studies, not only for the identification of geological hazards but also for collecting the information, was obtained by the Andean inhabitants in their permanent interaction with the Earth processes. In this context, it must be considered many important populations in the mountain region of Peru such as Cusco, Puno, Apurímac, Ayacucho, Huánuco and Cajamarca, which are growing and are demanding new lands for their urban, industrial and economic development.

Planning can reduce urban inequalities and restore self-regard to Peruvian cities (Bankoff, 2017; Macedo et al. 2014; Practical actions, 2017). The loss of human life and property produced in Peru in 2017 could be reduced if authorities had applied geotechnical designs for construction of houses for city dwellers, based on geological investigations. To help with this task, the participation of the geoscientist community in Peru is crucial in the management of socioenvironmental problems linked to mining activities, land planning and corruption. As pointed out succinctly by Desikachari (2015), since corruption plays a negative role in formulating geologic research in developing countries, combating this using a forceful geoethical approach by geoscientists becomes vital. Strengthening the social implications and impact of geosciences promotes a wide discussion among colleagues and the society about the role of geoscientists in developing an ethical approach to problems affecting the population.

2 Natural Hazards in Peru

The territory of Peru is located in the western centre of South America and all its territory, from South to North is crossed by the Andes Mountain Range, a characteristic landform of the country and closely linked to the development of Peruvian society.

The historical background of Peru shows it as a country whose population has permanently lived with geological and climatic hazards, facing them not to challenge the difficult environment, but to adapt to it. Some of the main natural hazards that have been affecting the social and economic development of Peru are:

- Mass movements
- Volcanic eruptions
- Glacier detachments
- Floods
- Earthquakes
- Tsunamis.

For each of them, there are examples of enormous magnitude that have affected a large part of Peruvian society, some of them even with a global scope, such as the eruption of the Huaynaputina volcano in 1600, many of which have not only generated the loss of human lives, material destruction, but also the destruction of the means of production that have caused thousands of injuries, deaths and billions of dollars in reconstruction work to replace damaged infrastructure and to attend the population affected by these geological hazards and recover urban areas, agricultural land, mining, agroindustrial and industrial areas.

The archaeologist Ruth Shady, who researched the Sacred City of Caral, the oldest culture in Peru (3000 BC–1800 BC) and one of the six worldwide areas, where the indigenous civilizations took place, explains about the Peruvian situation (Shady & Christopher, 2015). "The great geographic diversity of the central Andean territory, markedly contrasted in altitude, latitude and geomorphology, has promoted a peculiar adaptation process which has been experienced throughout these millennia. The instability of the geographical conditions, the periodic warming of the marine waters, changes in the sea level, tsunamis, tectonic activity, droughts, floods, frosts, the Puna's cooling, among others, was assumed in that adaptation through the use of different environments and the development of mixed economic activities".

It is believed that the Caral civilisation was affected by the climate system variability in the Andean zone due to the climatic phenomena known as El Niño-Southern Oscillation (ENSO, Fig. 2), the surface temperatures of the tropical Pacific and the Atlantic and other factors whose research are in progress (Perry et al., 2014; Villacorta et al., 2019). It affects Peru and the coastal region of South America (in the intertropical and equatorial zones), and its origin is related to the warming of the eastern equatorial Pacific Ocean. This phenomenon, in its most intense manifestations, causes heavy rains, which produce activation of floods and river overflows.

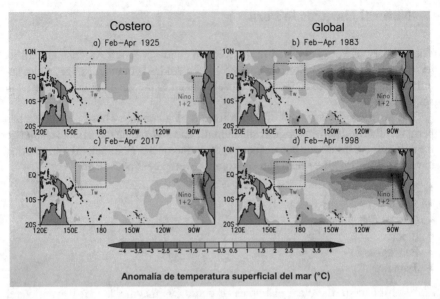

Fig. 2 Temperature anomalies related to El Niño-Southern Oscillation (ENSO). Technical Bulletin of Geophysical Institute of Peru (IGP) Vol.4, April 2017

For instance, according to Huertas, (1987), ENSO produced the activation of debris flows, which caused the collapse of the agricultural economy in the north of Peru.

The 1925 ENSO is used as a reference and the first graphic records photographs by Enrique Brüning (Schaedel, 1988, 1990), showing the disaster generated in the towns of Piura, Lambayeque and La Libertad, all located on the coast of northern Peru. In 1983, other ENSO events produced a great magnitude of rainfalls, which affected the populations of Tumbes, Piura, Lambayeque, La Libertad and Ancash more seriously. Most of the cities of the northern coast of Peru were affected by floods and debris flows. Channels, bridges and roads were destroyed, in addition to that, agricultural production was affected. It took too many points of the Gross Domestic Product to recover from that catastrophe and a delay in the economic and social development. Subsequently, there were two more events of ENSO, equally disastrous in 1998 and 2017, which has led researchers to take into account that the increase in the frequency of this type of climate phenomenon is probably influenced by the climate change. In the case of Peru is causing the warming of the waters of the Pacific Ocean and the subsequent impact on their territory. The 2017 Coastal El Niño led the Peruvian government to consider a budget of almost 7.5 billion dollars for reconstruction of the affected areas (Rodríguez-Morata, 2019). According to the damage assessment report of the Peruvian Civil Defence Institute (INDECI, 2017), these events produced in Lima 1,758 collapsed homes, 916 uninhabitable and 10,250 affected in 2017.

According to researcher Walter Alva (1986) "between 1800 BC and 1300 BC, in the area of Purulen (Lambayeque), exists the first vestige of human development with ceramics, which abandoned the zone due to the intensity of ENSO. The area comprised a set of 16 platform temples and about three-square kilometres of settlements that would be one of the first urban centres of the Lambayeque Region (North of Peru). At the moment, an investigation is being developed that allows to have many answers to the question about the first events of ENSO which probably occurred 1300 years before Christ".

Another of the great pre-Columbian civilizations of ancient Peru is the Mochica Culture (Moche, Sipan and Sican), which developed on the north coast of Peru between what is the coast of the current regions of La Libertad and Lambayeque, between 150 AD and 800 AD. Several theories try to explain its disappearance. Most agree that it is linked to the effects of ENSO, with its aftermath of floods and mud flows.

The researcher Marco Rosas Dintel (2007) states the following: "The argument of the environmental disturbances as triggering factor for Moche collapse arises from studies made between 1979 and 1980 in glaciers of the Andean mountains (Thompson, 1980; Thompson & Mosley-Thompson, 1987; also see Thompson et al., 1979; 1984; 1985; 1986).

Ice samples obtained from the deep drilling of glaciers made it possible to assemble a detailed 1500-year rainfall record. This record shows an environmental instability pattern during almost the entire extension of the late Moche period (ca. 550–800 D.C.), characterized by the succession of several severe droughts and interrupted by a single episode of excessive environmental humidity (Shimada et al., 1991: 261; Shimada, 1995: 249). Evidently, this period of climatic instability would have affected the productive capacity of the Moche communities settled on the coast. A point was probably reached, where the productive base of society was unable to provide the resources demanded by the ruling class. This would have generated the inevitable collapse of that society.

The Inca Empire (1200 AD–1533 AD) developed its population settlements in the Andes Mountain Range mainly, between the heights of 2000 and 3500 m above sea level. The vestiges of its citadels remain today as a monumental heritage of humanity. Examples are located in Cusco, Huanuco, Cajamarca and the archaeological monuments of Machu Picchu, Ollantaytambo, Sacsayhuaman, Tipon, Pisac, Choquequirao, among others. All these constructions show an adaptation to complex environmental conditions such as heavy rains, steep slopes, crumbling soils and others that were controlled by Inca engineering.

Likewise, the Inca Empire has left, as a heritage, a network of paths and bridges of more than 60,000 kms known as QhapaqÑan. This was organized on a continental scale through the current territories of Peru, Ecuador, Colombia, Chile, Bolivia and Argentina (González-Godoy, 2017). After 500 years of carelessness, QhapaqÑan continues being used by various indigenous communities, without maintenance or improvement. Most of its roads have remained as monuments and vestiges of the engineering work of adaptation to nature carried out by the Incas. The Inca Road is a sharp contrast to the current roads and highways in Peru, which, when not constantly maintained, collapse. Due to mass movements such as landslides, rock falls, mudflows, debris flows, gullies, erosion, etc., which are not well controlled by the current engineering, it is produced constant interruption of Peruvian roads and also infrastructures such as bridges, channels, hydroelectric plants, buildings and others. Those processes continuously have been severely affecting various popula-tions settled mainly in the Andes Mountain Range. In reason of the great damage that these processes caused to the Peruvian economy, Villacorta et al. (2012) analysed the susceptibility to landslides in Peru and their results concluded that the areas with the highest probability of being affected are mainly located in: (1) in the western part of the Peruvian Andes, in the area of Cajamarca-La Libertad-Ancash-Lima-Huancavelica; (2) in the south-western area (Ayacucho-Apurimac-Cusco-Puno); (3) in the south-eastern area (Arequipa-Moquegua-Tacna); (4) In the central and north-eastern region (Junín-Pasco-Huanuco-San Martín) (Fig. 3).

Among the most important events can be mentioned the landslides that buried the Machu Picchu Hydroelectric Plant and part of the town of Santa Teresa in 1997 and the avalanche of Mirave (Tacna) in 2017. However, the higher level of risk is located in Lima, mostly on the slopes of the middle and upper basins of the Chillon, Rimac and Lurin rivers because of the highest concentration of population (Villacorta et al., 2015). Chosica (Lurigancho District, Lima) is an example of populations most-at-risk in Peru. The town is located on unstable slopes where rockfalls and debris flows are likely to happen and close to the Rimac River, which has a high risk of damaging and destructive flooding. For these reasons, it is constantly affected by large debris flows registered since 1907. Those events caused deaths, extensive property damage and disruption to the economy and well-being of Lima's community in 1982, 1998, 2012, 2015 and 2017 (Villacorta et al., 2020).

Other important geological risks that affect the population of the South of Peru are those related to Volcanism. According to Thouret et al. (2002), in the year 1600 AD, Huaynaputina volcano located in the southern Peru (Moquegua), presented a large Plinian-type eruption, with a Volcanic Explosivity Index 6 (VEI 6), that is considered the largest eruption that occurred in South America (Fig. 4) in historical times. The eruption caused the death of approximately 1500 people, burying at least 11 villages, which were located less than 20 km from the volcano.

The successive reactivation of two volcanoes in southern Peru, the snowy Saban-caya in Arequipa (between 1987 and 1998), put the Majes-Siguas canal, a source of water for 35,000 people, at risk. Subsequently, the Ubinas volcano (Moquegua) erupted in 2006, forcing 1,500 people to evacuate in 2014, and then in 2019, it resumed its eruptive activity, forcing the permanent relocation of the town of Querapi, located close to it. It is worth noting that the area was declared in 2014 as a very high risk that cannot be mitigated, so the area must remain inhabited.

One of the most recurrent geological hazards in the territory of Peru and which throughout the years have affected millions of Peruvians are the earthquakes (Fig. 5), whose historical records are included in the chronicles of the stage of the Viceroyalty of Peru (1533–1824), Cobo (1653), when Peru was a colony of Spain.

The first information about a big earthquake goes back to the time of the Tupac Yupanqui Inca (Silgado 1973), then from 1533 to date, other devastating earthquakes have struck especially the cities of southern Peru such as Arequipa, Tacna, Moquegua, Caravelí, Nazca, Ica and Pisco. Even Lima has suffered devastating earthquakes like

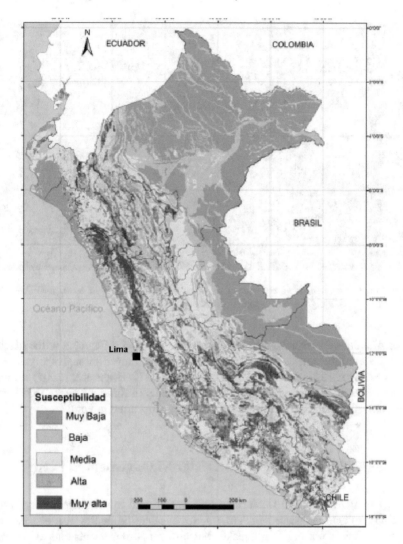

Fig. 3 Mass movement susceptibility map of Peru (Villacorta et al., 2012). In red colour the areas with higher levels of susceptibility to these processes

the ones in 1650, 1746, 1940 and the last, less intense, in 1974. Now, there is a worrying seismic gap of almost 50 years.

The earthquake in 1746 destroyed 99% of the houses in Lima and 95% of people living in Callao (the Lima harbour), including the entire population of the aboriginal fishing community known as Piti Piti, died, due to the subsequent Tsunami.

The earthquake of 1970 with a 7.7 magnitude, on the coast of Chimbote (Ancash region, located at 400 km to the north of Lima), was one of the most devastating in Peru's history. 70,000 people died; 25,000 of them in the Yungay village, due to the

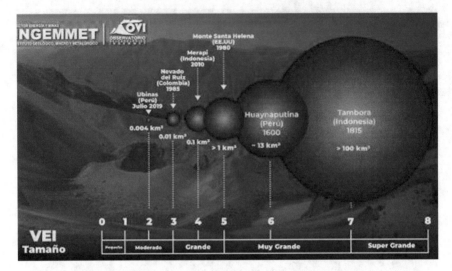

Fig. 4 Magnitude of Volcanic eruption of the Huaynaputina Volcano in 1600 (INGEMMET), compared with other big volcanic eruptions occurred in historic times

action of a debris avalanche caused by the detachment of a fragment of the glacier which covered the Huascaran glacier. The avalanche moved through the valley at speeds close to 270 km / h (Fig. 6), (Ericksen et al., 1970; Evans et al., 2009) burying people under 30 meters of mud and rock fragments of different sizes. That is the most catastrophic internationally known event that occurred in Peru.

3 Stakeholders Involved in Disaster Management in Peru

Starting in 1970, the Civil Defence system was created to work adequately in the prevention of natural disasters in Peru, a series of scientific studies were also carried out, to catalogue not only earthquakes but also mass movements and to monitor glaciers evolution. Nowadays, several institutions work in coordination as part of a permanent National Emergency Operations Centre.

There are scientific institutions such as the Geological, Mining and Metallurgical Institute (INGEMMET), Geophysical Institute of Peru (IGP), Japanese-Peruvian Centre for Seismic Research and Disaster Mitigation (CISMID) of the National University of Engineering, and disaster management governmental bodies, such as the National Institute of Civil Defence (INDECI), and the National Centre for Estimation, Prevention and Reduction of Disaster Risk (CENEPRED), which for a few years, have been working in coordination to confront, prevent and mitigate disaster risk in the country.

The Volcanological Observatory of INGEMMET (OVI) operates from the city of Arequipa and is a centre for the study and monitoring of active volcanoes in southern

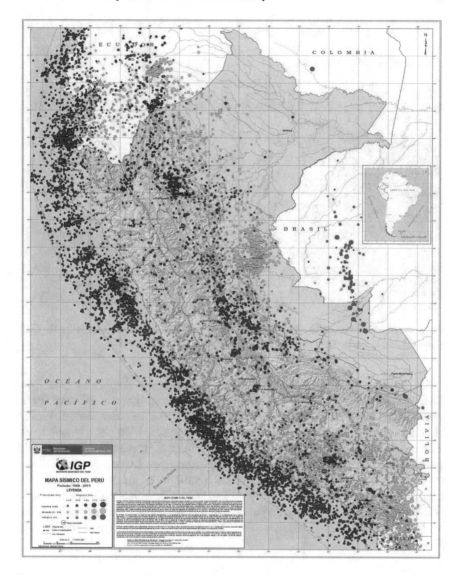

Fig. 5 Sismic map of Peru (IGP. 2019)

Peru. OVI evaluates the types of volcanic hazards based on geological studies and specialised analysis and provides timely alerts to Peruvian citizens, who live in the south of the country, about imminent volcanic activity, to reduce the risk of disasters. It monitors the central volcanic zone of the Andes (CVZ), where there are 12 active volcanoes and potentially active volcanoes of southern Peru. There are at least seven regions (Sabancaya, Misti, Ubinas, Huaynaputina, Ticsani, Yucamane, Tutupaca) that have presented eruptive activity in the last 500 years (Sieber et al., 2010).

Fig. 6 Aerial view of the 1970 Avalanche, which buried the town of Yungay and caused the death of 25,000 people (taken from: http://www.actualidadambiental.pe/?p=19482)

On the other hand, in 2004, in Caracas (Venezuela), geological services from Andean countries (Argentina, Bolivia, Canada, Chile, Colombia, Ecuador, Peru, Venezuela) and Canada met to work on an institutional initiative, called Andean Multinational Project: Geosciences for Andean communities, the result of which was the creation of GEMMA (Group of Standards for Mass Movements), who developed a series of publications (Fig. 7), which have become a guide for our countries and have helped to have an Andean vision of the occurrence of the diverse mass movements that permanently affect the region and whose investigation is necessary, in order not only to understand them but also to propose prevention and socialization work with the inhabitants and authorities, to mitigate the destructive effects of these phenomena.

4   Risk Perception and Preparedness

The unknowing of the particular geodynamic configuration of Peru and its constant evolution (Villacorta et al., 2012) by local authorities are linked to mismanagement to deal with the settlement without or with ineffective planning. Authorities are not well prepared to face the massive demand for housing by Peru's migrant population to coastal capitals of Peru. Under their pressure, the Peruvian authorities, unaware of how to deal with disasters in the country, have allowed the informal occupation of

Fig. 7 Hydraulic Risk Map produced under the Pilot Project of Andean Multinational Project: Geosciences for Andean communities. INGEMMET 2004

lands and the granting of property titles in areas prone to geological hazards. This has only increased the risk of vulnerable populations and indicates a lack of knowledge about the geosciences' usefulness to solve this kind of situation.

In most cases, planning laws are not understood by the population and consequently cannot be properly applied. Therefore, decisions and actions are not based on scientific information or from a multidisciplinary perspective. Geoscientists are called only as observers in the resettlement process of high-risk populations (Macedo et al., 2014). Taking into account that Peruvian cities do not have enough empty lands for more people to settle, it is evident and of the utmost importance to take measures for its proper management.

Furthermore, Peruvian citizens are not accustomed to using geoscientific information even when it is available. This is one of the causes of their late reaction to disasters. In many cases, after those disasters occurred, houses continued being built in high-risk areas without adequate technical support.

5 Discussion and Geoethical Implications

Without a scientifically prepared society, effective and widespread risk reduction strategies cannot be developed. The result will be the development of an emergency culture rather than one of risk preventions, with an increase in victims and repercussions of disasters on the next generations (Vasconcelos et al., 2020). Being part of human behaviour, reflect individually and collectively to correct and repair actions (Castoriadis, 1994); by explaining the complex geological and climatic conditions of Peru, the attitude of citizens towards disaster prevention can be modified. For example, explaining the risks of clogging river channels with rubbish will help the community actions towards avoiding it.

Identifying the relevant stakeholders that geoscientists should work with, and using appropriate channels to communicate scientific information on georisks to the public, can support the efficient incorporation of the principle of prevention. This establishes that in the face of irreversible environmental damage, or whose reversibility is only achieved with enormous effort, preventive measures are justified even in the absence of conclusive scientific evidence (Almeida, 2014). The geoscientist should contribute efficiently to this principle by providing citizens with information on scenarios that allow risk reduction. In this, it is useful to consider what is proposed by Frodeman (2004), about imagining an era in which technological progress slows down to respect natural limits, as a way of preventing critical scenarios for humanity.

At the governmental level, given the high impact and low predictability of georisks, the development of a priori risk management is mandatory (Almeida, 2014). In this context, the implementation in short time of inexpensive and low-maintenance solutions are necessary. For instance, declaring unsafe places and prohibiting the reconstruction of new homes there, unless scientific studies provide an adequate understanding of the disaster causes and how to reduce the risk (Villacorta et al., 2020).

Willingness on the part of national agencies such as INDECI and CENEPRED is needed to provide support for weak areas such as specialist training and providing outreach materials. To implement this initiative, those entities must work in coordination with the other Peruvian institutions that are part of the National Disaster Risk Management System.

To facilitate the population in adapting to the difficult environmental conditions in Peru, it is crucial to update the laws of land use and management (infrastructure, public space, transport, etc.) as a guide to plan the orderly growth of cities. This planning should consider the conclusions and recommendations of scientific studies, especially for the reconstruction of communities affected by georisks. Geoscientists of the mountain region should contribute to researching the geological conditions of the territories located in this part of the country as well as their vulnerability and risk because those are required for the development of cities located there. In these locations, it could be considered to recover the use of ancient construction techniques such those applied by The Incas in Machu Picchu citadel. It has proven to be effective in dealing with mass movements and is earthquake-resistant structures (Cuadra et al., 2005).

Furthermore, raising funds for scientific research and science communication is also crucial. It is fundamental to invest in catalogue and study not only the most recent events but also the old ones and increasing the number of meteorological stations. It should be taken into account, of course, extreme climatic events such as ENSO and other climatic anomalies. That allows forecasting more accurately those kinds of events in the different regions of Peru.

6 Conclusions

After this overview, it is possible to understand that Peruvian communities have been living for almost 6000 years with many geological and climatic events, searching for a continuous adaptation to the catastrophic effects of natural phenomena. However, some circumstances influence this adaptation and generate a high risk to the populations. In particular, economic and social factors due to the urban improvisation on the part of national authorities have led the inhabitants to concentrate in large cities, mainly in the coastal region of Peru. Informality, speculation, inability and mismanagement of local authorities are also big problems in Peru.

Populations are exposed to the catastrophic action of earthquakes (due to vulnerable self-constructions, with little territorial planning and minimal technical direction), water scarcity (most important cities are located in desert areas), pollution and geo-hydrological processes.

It is clear that, in Peru, geo-hydrological processes such as floods and debris flows (produced by phenomena such as ENSO) will always occur and affect populations settled along river's channels and banks due to insufficient knowledge, negligence or omission.

In this context, it has been shown that it is still required to develop management policies, land use planning and scientific research to understand geological hazards. It contributes to reducing the possibilities of damage by disasters, increasing resilience and improving the quality of life of Peruvian communities. This will result in an improvement in the economy that is regularly affected not only by the interruption of economic activities but also by the material damage due to disasters.

To raise these goals, geoscientist community participation using a geoethical approach is crucial.

References

Almeida, A. (2014). A Geoética no Currículo e na Formação de Professores. *GEOlogos, 11.* Retrieved from https://repositorio.ipl.pt/bitstream/10400.21/11366/1/A%20Geo%C3%A9tica%20no%20Curr%C3%ADculo%20e%20na%20Forma%C3%A7%C3%A3o%20de%20Professores.pdf.

Alva, W. (1986). Investigacionesen el complejoformativo con arquitectura monumental de Purulén, costa norte del Perú/ Untersuchungen in demformativzeitlichenKomplexmitMonumentalarchitektur von Purulén, NordküstePerus, BeiträgezurAllgemeinen und VergleichendenArchäologie 8, 283–300, Mainz am Rhein.

Bankoff, G. (2017). Living with hazard: Disaster subcultures, disaster cultures and risk-mitigating strategies. In *Historical disaster experiences* (pp. 45–59). Springer International Publishing.

Castoriadis, C. (1994). The misery of current ethics. *Letrainternacional.* ISSN 0213-4721, N° 32.27–33

Cobo, B. (1653). Historia del Nuevo Mundo, crónicas de Perú, México y el Caribe.

Cuadra, C., Sato, Y., Tokeshi, J., Kanno, H., Ogawa, J., Karkee, M. B., & Rojas, J. (2005). Evaluation of the dynamic characteristics of typical Inca heritage structures in Machupicchu. *WIT Transactions on The Built Environment, 83.*

Desikachari, V. (2015, April). The relevance of geoethics to under-developed and developing Nations with special reference to India. I. In *EGU General Assembly Conference Abstracts* (Vol. 17).

Ericksen, G. E., Plafker, G., & Concha, J. (1970). *Preliminary report on the geologic events associated with the May 31, 1970, Peru earthquake* (36 p.). Geological Survey Circular 639. Washington: US Geological Survey.

Evans, S. G., Bishop, N. F., Smoll, L., Murillo, P., Delaney, K. B., & Oliver-Smith, A. (2009). A re-examination of the mechanism and human impact of catastrophic mass flows originating on Nevado Huascaran, Cordillera Blanca, Peru in 1962 and 1970. *Engineering Geology, 108*(1), 96–118.

Frodeman, R. (2004). Philosophy in the field. In B. W. Foltz, B. V. Foltz, & R. Frodeman (Eds.), *Rethinking nature: Essays in environmental philosophy* (pp. 149–164). Indiana University Press.

González-Godoy, C. (2017). Road archeology of the QhapaqÑan in South America: Theoretical analysis, concepts and definitions. *Bulletin of the Chilean Museum of Pre-Columbian Art, 22*(1), 15–34. In Spanish.

Huertas, L. (1987). Ecología e Historia. Probanzas de indios y españoles referentes a las catastróficas lluvias de 1578, en los corregimientos de Trujillo y Zaña. Centro de Estudios Sociales "Solidaridad", Chiclayo.

INEI Census. (2017). www.censo2017.inei.gob.pe.redatam.

Instituto Nacional de Defensa Civil-INDECI (2017). Compendio Estadístico del INDECI, 2017 – Gestión Reactiva/ Perú. Lima: INDECI. DIRECCIÓN DE POLÍTICAS, PLANES Y EVALUACIÓN, 2017. 290 p.

MaedaAscensio, J. (2015). "El Niño": Análisis Histórico y su impacto en el ambiente natural y en las culturas prehispánicas, hasta la actualidad. Retrieved from http://josemaeda.wordpress.com/ 2015/11.

Marco Rosas Dintel (2007) New perspectives on the Moche collapse in the lower. *Jequetepeque - Bulletin of 'Institute francaisd´etudesandines, 36*(2), 221–240

Rodríguez-Morata, C., Díaz, H. F., Ballesteros-Canovas, J. A., Rohrer, M., & Stoffel, M. (2019). The anomalous 2017 coastal El Niño event in Peru. *Climate Dynamics, 52*(9–10), 5605–5622.

Practical Actions. (2017). Peru 2017 Risks, Disasters and Reconstruction.

Perry, L. B., Seimon, A., & Kelly, G. M. (2014). Precipitation delivery in the tropical high Andes of southern Peru: New findings and palaeoclimatic implications. *International Journal of Climatology, 34,* 197–215.

Schaedel, R. P. (1988). La etnografía muchik en las fotografías de H. Brüning, 1886–1925. Lima: Ediciones COFIDE.

Schaedel, R. P. (1990). El legado de Brüning – su redescubrimiento. In: Raddatz, C. (Ed.): Documentos fotográficos del norte del Perú (pp. 36–38). Hamburg: Hamburgisches Museum fürVölkerkunde.

Shady, R., & Christopher, K. (2015). Caral. La primera civilización de América. The first civilization in the Americas, Ed. San Martín de Porres University.

Shimada, et al. (1991). Investigation of prehistoric copper production at Batán Grande, Perú: Interpretation of the Analitical Data for Ore Samples. *Paper presented at the International of the Archaeometric Symposium, The Getty Conservation Institute, Marina del Rey, CA.*

Shimada (1995). Ground penetrating radar: Large scale application on coastal Peru. In *Paper presented at the 60th Annual Meeting of the Society for American Archaeology, Minneapolis.*

Thouret, J. C., Juvigné, E., Gourgaud, A., Boivin, P., & Dávila, J. (2002). Reconstruction of the AD 1600 Huaynaputina eruption based on the correlation of geologic evidence with early Spanish chronicles. *Journal of Volcanology and GeothermalResearch, 115*(3–4), 529–570.

Thouret et al. (2018). Cronoestratigrafía del volcanismo con enfásis en ignimbritas desde hace 25 Ma en el SO del Perú – Implicaciones para la evolución de los Andes Centrales. En Foro Internacional: Los volcanes y su impacto, 8, Arequipa, PE, 26–27 abril, 2018, Libro de resúmenes. Arequipa: INGEMMET (pp. 36–42)

Thompson, L. G. (1980). Glaciological investigation of the tropical Quelccaya ice cap, Peru. *Journal of* **Glaciology**, *25*(91), 69–84.

Thompson, L. G., & Mosley-Thompson, E. (1987). Evidence of abrupt climatic during the last 1,500 years recorded in ice cores from the tropical Quelccaya ice cap, Peru. In: W. Berger (Ed.) *Abrupt climatic change – evidence and implications* (pp. 99–110). D. Reidel Publishing Co.

Thompson, L. G., Mosley-Thompson, E.., & Petit, J. R. (1979). Glaciological interpretation of microparticle concentrations from the French 905-m Dome C, Antarctica core. In I. Allison (Ed.), *Sea Level, Ice and Climatic Change: Proceedings of the Canberra Symposium, December 1979* (Vol. 131, pp. 227–234). IAHS Publication.

Thompson, et al. (1984). Major El Niño-Southern Oscillation events recorded in stratigraphy of the tropical Quelccaya Ice Cap. *Science, 226*(4670), 50–52.

Thompson, et al. (1985). A 1500 year record of climate variability recorded in ice cores from the tropical Quelccaya Ice Cap. *Science, 229*(471), 971–973.

Thompson, et al. (1986). The "little Ice Age" asd recorded in the stratigraphy of the tropical Quelccaya ice cap. *Science, 234*, 361–364.

Vasconcelos, C., Schneider-Voß, S., & Peppoloni, S. (Eds.) (2020). Enseñanza de Geoética. Recursos para la educación superior. U. PortoEdições, 207 pp. https://doi.org/10.24840/978-989-746-254-2.

Villacorta, S. P., Fidel, L., & Zavala, B. (2012). Mass movement susceptibility map of Peru. *Revista De La Asociación Geológica Argentina, 69*(3), 2012.

Villacorta, S. P., Nunez, S., Tatard, L., Pari, W., & Fidel, L. (2015). Geological hazards in the Metropolitan Lima area and the Callao Region (Lima, Peru). *Bulletin of the Geological, Mining and Metallurgical Institute, Lima, 151*, 59.

Villacorta, S. P., Evans, K. G., De Torres, T. J., Llorente, M., & Prendes, N. (2019). Geomorphological evolution of the Rimac River's alluvial fan, Lima Peru. *Geosciences Journal, 23*(3), 409–424. https://doi.org/10.1007/s12303-018-0049-5

Villacorta, S. P., Evans, K. G., Nakatani, K., & Villanueva, I. (2020). Large debris flows in Chosica, Lima, Peru: The application of hydraulic infrastructure for erosion control and disaster prevention. *Australian Journal of Earth Sciences, 67*(3), 425–436.

Volcanological Observatory of INGEMMET-OVI (2020). *What is the INGEMMET volcanological observatory?* Retrieved from http://ovi.ingemmet.gob.pe/.

Geoethics and Mining Activity in Peru

**Esteban Manrique Zuñiga, Carlos Toledo Gutierrez,
and Sandra Paula Villacorta Chambi**

1 Introduction

Mining has been considered an unsustainable activity for many years because it includes the physical removal from the ground of a non-renewable resource for further processing and use. The sustainability of extractive activities is a relatively recent concept, which includes geoethical values towards the community involved in extractive projects (Limaye, 2012).

The development of a country is reflected in the advancement of science and technology. In Peru, this technological and industrial development is linked to the exploration and mining of minerals. In this context, the geoscientist has multiple roles. Firstly, in academic research and secondly, interacting with nature and the world. The main objective is to raise the profile of geoscientific knowledge in the public and educate the uninformed population based on a foundation of geoethical values and resulting in an ethical attitude towards the sustainable development of a country.

Some numerous geoethical concerns evident in Peru are related to a fast-growing economy associated with mining activity. Peru is globally well located as a region for

Electronic supplementary material The online version of this chapter
(https://doi.org/10.1007/978-3-030-86731-7_4) contains supplementary material, which is
available to authorized users.

E. Manrique Zuñiga (✉)
National University of Engineering, Rímac, Peru
e-mail: emanrique@uni.edu.com

C. Toledo Gutierrez
Antonio Ruiz de Montoya University, Pueblo Libre, Peru
e-mail: carlos.toledo@uarm.pe

S. P. Villacorta Chambi
Charles Darwin University, Casuarina, Australia
e-mail: sandra.villacorta-chambi@cdu.edu.au

mineral exploration because of its abundant reserves of economically viable minerals (APCSG, 2011; Garcia, 2011). The ethical problem involved in Peruvian extractive activities is exemplified by the uncontrolled mineral exploitation in some Peruvian regions (Gardner, 2012; Langeland, 2015). In these regions there is social conflict involving farmer communities which continue to protest against the contamination of, and demanding protection for, fluvial systems (Li, 2013, 2016).

A recurring problem for mineral exploration projects is obtaining a social licence to operate. The biggest issue is the lack of dialogue about the extraction and use of any natural resources between an uninformed population and entrepreneurs who do not apply geoethical values (Di Capua et al., 2018). Then, misunderstandings are capitalized upon by those who wish to generate chaos and social conflict. However, in Peru, the ethical problem involves the Peruvian government decision to allow mineral resources to be extracted without strict enforcement of state regulation, which has led to uncontrolled extractive activities in some Peruvian regions (Gardner, 2012; Langeland, 2015).

In this framework, the use of geoethics is particularly useful to avoid potential conflict and is especially relevant in Peru, where there is a need to raise awareness about the effects of exploring and developing extractive activities. This will allow proposing alternative solutions to promote responsible extractive activities while taking care of the environment.

2 Mining Activities in Peru

Mineral resources have been exploited in Peru since pre-Inca times, becoming more widespread in the colonial and the republican era. There has been a notable increase in mining activity in the last three decades due to significant national and foreign investment that has led to a profitable Peruvian industry (Durand, 2016). For example, the Yanacocha mine in Cajamarca is a world-class gold and silver mine, is one of the most important mines in the country and was fully developed by Peruvian professionals with foreign capital.

Mineral explorers carry out their activities on private lands (individual or communal). If they have not sought permission to access the land, owners are upset when they find them on their properties. Thus, the problem stems from the explorer's attitude towards the owners, as well as the ignorance of the landowners, communities and local authorities about environmental protection laws and the contribution of mining to the country's economy. Sharing this information is a task for both the State and the companies involved in the extractive activities.

As a result of the lack of effective communication and proactive citizen participation, environmental proposals and commitments offered by mining companies to Peruvian communities are short-lived. In other words, there is no continuity in the efforts to solve any environmental and social responsibility issues by the actors involved in those projects. It should be emphasized that both social and environmental

Fig. 1 A field worker learning to recognize different types of quartz. Photo E Manrique 2011

responsibility are equally shared between extractive companies, communities and the Peruvian State.

Extractive companies have responsibilities in the field for guaranteeing fair wages and social benefits for each worker, according to their level of responsibility and job description. Raising of the company's profile through positive social engagement is desirable on the part of the company, but should not be confused with basic welfare that seeks to replace the duties of the Peruvian State towards society. Environmental responsibility must be expressed through sharing of basic knowledge about effective environmental remediation technologies, and seeking to train community residents interested in learning how to use or provide assistance in the application of these remediation techniques.

Extractive activities are also an opportunity for the villagers to train in exploration tasks (Fig. 1) and often become valuable field support for specialists.

3 Community Perception

Knowledge of the scope about any mining or exploration activities is vital for the inhabitants. The best way to overcome disinformation is to educate the population, a task of the state, and the extractive companies through their departments of community relations. Education has to be comprehensive and begin in the primary stages of education. It is essential to work on raising awareness in society about caring for the environment and with extractive companies to mitigate or remediate the impacts of their activities. For instance, although mining environmental liabilities (PAM) are still considered a threat to the residents living in its surroundings, most of them are unaware of the current norms related to the remediation of the environment after mining.

The residents of the communities are easily manipulated by ideologized people with anti-mining positions who are keen to take advantage of extractive companies through association of those activities with death. In their view, environmental and natural resource pollution creates life-threatening conditions. For this reason, the anti-mining discourse uses slogans in defence of life. Until this ideological problem is addressed and resolved, communities will believe that the labour and material benefits offered by extractive companies are of no use if ultimately "everyone is going to die from high levels of contamination." It is evident that good promises are useless in stopping an anti-large-scale mining attack. An enlightened democratic dialogue strategy is a good start where ideas are fought with ideas and arguments with better arguments. This is one of the reasons for the failure of community relations specialists. Environmental philosophers and pedagogues are needed to discuss the arguments against mining, as well as anthropologists and sociologists.

4   Regulatory Governmental Entities in the Decision-Making Process

Exploration is the first step in a mining project and formal activities are regulated through the General Mining Law or Supreme Decree N° 014-92-EM (MEM, 2020) and environmental regulations issued by the Ministry of the Environment. In Peru, the extractive activity implies concession of the land, which means obtaining a social licence to commence exploration of the land.

At the cessation of mining, PAM, which includes tailings produced by abandoned mining operations and currently inactive, effluents and emissions, constitutes a permanent and potential risk for the health of the population, the ecosystem and the property. In Peru, these deposits are regulated by Law 28271 (2004), however, in many cases, it is difficult to mitigate the risks produced by abandoned mining to the environment.

Furthermore, even though the Peruvian Government has established a legal framework of maximum permissible limits for mining development, it has not been possible to halt some social conflict. Farmers and rural people continue denouncing and protesting against the expansion of the mining industry, highlighting the contamination of fluvial systems and claiming the need for protection of water resources (Li, 2013, 2016).

The national government and corresponding state institutions agreed on a policy to support the minerals industry by promoting the extraction and export of raw materials. This legislation seeks to create favourable legal conditions for foreign investment and mixed capital expenditure which develop extractive operations in Peruvian territory. However, this does not solve the current socio-environmental conflicts. In a democratic context, discussion and antagonistic positions are common, since not all citizens can be expected to agree. In Peru, the idea of maintaining a democratic dialogue in search of a consensual solution is complicated and can lead to violence. Consequently, the activities or investments of some extractive projects have been stopped.

5 Geoethics Applied to the Mining Industry

Communication between explorers and landowners involved in the exploration of mineral deposits is of the utmost importance. The success or failure of the project depends on that dialogue (Fig. 2). Dialogue must be sincere and timely to establish a good relationship between both parties and to allow the successful start of the activity. In the communication process, the work to be carried out by exploration geologists must be presented simply and clearly. In this task, geoethics is important because it can be applied to explain the potential benefits and disadvantages of exploration. Activities to be carried out in the field, such as sampling, digging of

Fig. 2 A sincere dialogue with the owners of the land to be explored is crucial. Photo E Manrique 2011

Fig. 3 Teamwork discussing with community members to open trenches and exploratory pits. Photo E Manrique 2011

wells and/or trenches, construction of accesses and camps (Fig. 3), must be specified (Boon, 2020). All these activities generate an environmental impact and a serious commitment must be assumed for their remediation. In mining, there is no bad luck, only bad decisions or trying to solve problems without applying professional ethics. Consequently, social and economic issues arise that are detrimental to the country's development.

In this framework, geoscientists should research techniques for the development of exploration and mining, reducing the possibility of impacts on the environment. They must operate reasonably and make responsible use of mineral resources concerning the real needs of society. Sustainable practices must always be adopted and a geoscientific approach used in their studies and decisions (Careddu et al., 2019).

Geoethics has an important role in sharing values to the personnel of the extractive companies, the environmental and social ethical reflection amongst staff from across the company, from the managers and professionals of the geosciences, to administrative staff and workers. Induction discussions should not only be related to occupational health and safety issues, but also to the importance of adopting geoethical values in the development of different tasks undertaken by the company. Geoethical reflection allows geoscientists to broaden their understanding of the true context of any scientific and professional activity. That is, incorporating the environmental and social context for the common good (Di Capua et al., 2018; Toledo, 2019).

6 Sustainable Mineral Production

Geoscientists are challenged with the intelligent management of ecosystems and natural resources for the benefit of the community. This challenge involves taking on new roles, professional skills, abilities and tasks that could be for an engineer, technician or geoscientist. Moreover, the implementation of geoethical values in professional development and practices implies that the care of life is the basis for consensus and solutions to socio-environmental conflicts. In other words, such conflicts should be assessed as new biopolitical scenarios (Abbott, 2020).

The development of technologies, that are able to remedy or recover areas where the environment is affected by pollution, is important for providing solutions to these issues and is of national and international interest. Companies and the government leaders should understand that investing in the research of new clean technologies and environmental remediation contributes to sustainable development and the optimization of natural resource extraction processes.

7 Discussion and Implications

A complex challenge in Peru is the process of obtaining permission to explore for minerals, which can last decades. Where is the problem? Who is negotiating? Unfortunately, the parties do not share a common goal and lack empathy for each other. The mining companies wish to explore on privately owned land, whether by individuals or as a collective (communities); whereas the landowners defend their property, claiming that mining will impact and contaminate the environment. The landowners are joined by interested parties who encourage them and obtain a political or economic benefit. This conflict between mining and environmental concerns has been occurring for many decades in Peru. The owners of the explored lands, who have based their opinions on traditional mining activities that have left enormous environmental liabilities, are quite right in this respect. They highlight that in the past, lands and natural resources were contaminated through mining activities that did not take into consideration care for the environment. Examples include the Ticapampa Tailings1 (Fig. 4) deposited in the late 1800s on the right bank of the Santa River (Recuay, Ancash, Peru) and the XVIII century tailing dams in the Rimac River Basin, where waste material was dumped into waterways of this Andean basin, which is located in the central part of Peru and is the main water resource of the Peruvian capital, Lima. Other Peruvian communities that have been affected by this kind of contamination include Oroya city, a metallurgical centre, where flora and fauna at the site were contaminated and exterminated. For this reason, international non-governmental organizations (NGOs), have been working in Peru for several decades to raise popular awareness about contamination (Arellano, 2010; Li, 2016). However, some Peruvian government representatives suggest that the problem is due to political opposition and the incompetent performance of local governments (Arellano,

Fig. 4 Ticapampa tailings deposited in the late 1800s on the left bank of the Santa River. Taken from: https://es.wikipedia.org/wiki/Relave_de_Ticapampa

2010). According to García (2007), the resistance of rural inhabitants to modernization process, which would bring new prosperity for Peru, is associated with an international conspiracy discouraging development of the country. This is supported by international NGOs and their national counterparts, the Catholic Church and radical groups which encourage an illiterate population towards a popular revolution (Arellano, 2010; García, 2007).

Other examples include the Conga mining project in Cajamarca, abandoned in 2016 after residents protested claiming that aquifers' contamination occurred at the head of the basin; Tía María in Arequipa, where farmers argue that there has been contamination of their crops; and artisanal and small-scale gold mining extraction in the Peruvian Amazon, where a rapid change in forest cover due to the expansion of agriculture and extractive activities has been reported (Nicolau et al., 2019). Despite the fact that contamination in Peruvian aquifers has also been linked to agricultural activity, domestic use and egress of industrial wastewater (Amezaga et al., 2007; Toledo, 2019), it is not possible to ignore the impact of mining activity. Unfortunately, there is limited research to correlate mining and the contamination of hydrological resources in Peru (Delman, 2012; Persson, 2009), which reflects poor management of the socio-environmental debate and arising conflicts by the mining companies and Peruvian State.

A new theoretical framework would allow these issues to be dealt with by building bridges of trustworthy dialogue between the three actors in conflict: the state, mining companies and communities. Perhaps biopolitics could be this new theoretical framework that initiates and maintains dialogue, since the value and importance of life is an undeniable topic of permanent discussion (Moore et al., 2015).

An example of long term environmental impact includes the tailing deposits at Ticapampa which has remained for hundreds of years as a source of pollution and shows that it takes a long time for natural processes to become inert. In these old, the alteration process generates acid solutions and toxic chemical elements pollute the water and exterminate the aquatic and terrestrial fauna. This type of environmental contamination continues today under artisanal and small-scale mining activities (Figs. 5 and 6).

Community members and leaders must also assume a social responsibility towards the environment. For example, they know how to work as a team with the company and local and regional authorities, and can collaborate to solve problems and any misunderstandings. Goodwill must prevail amongst all the stakeholders and requires to be coordinated work and dialogue for the common good. This proposal is consistent with ethics, prioritizes continuous improvement based on values, and is far from a modern false morality.

The Peruvian State and its authorities should be familiar with the theoretical framework of biopolitics, which is the strategy of liberalism to enhance the dynamics of production and the realization of conditions for the desired material culture. The main achievement of worker training is the concept of leadership and empowerment.

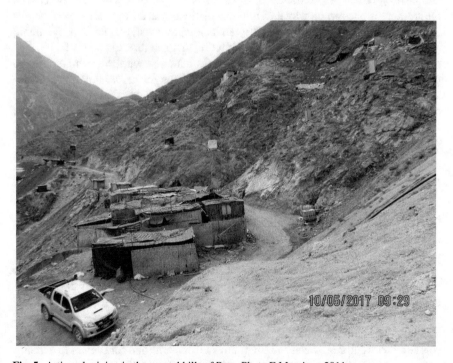

Fig. 5 Artisanal mining in the coastal hills of Peru. Photo E Manrique 2011

Fig. 6 Environmental pollution generated by artisanal miners. Photo E Manrique 2011

However, this aspect is unrelated to a State policy that not only seeks to empower the worker, but also improves their quality of life. This aspect involves the social and environmental responsibility of the state. The lack of a policy of life-care to the workers is one reason why anti-mining positions gain acceptance in populations (Toledo, 2019). By not feeling that they are being protected, people identify mining activities as a source of environmental pollution that can threaten their life.

State institutions must promote democratic dialogue with the communities to prevent physical violence. Implementing mechanisms for peaceful conflict resolution, such as negotiation, mediation and reconciliation, means that situations where complaints about human rights violations (which aggravates the problem and initiates legal procedures) can be avoided (Boon, 2020).

8   Conclusions

Overall, popularizing geoscientific knowledge will make it possible to promote a more favourable approach to extractive activities and improve its social acceptance: dialogue within an informed society will foster consensus and identify solutions.

In mining, there is no bad luck, only bad decisions or trying to solve problems without applying professional ethics. Any social and economic issues that do arise are detrimental to the long-term development of the country.

An aspect of great importance is that geoethics contributes to the development of favourable competencies for professional practice in geosciences. It promotes solidarity with cultural, social and environmental diversity and contributes to building a more fair and democratic society with a better quality of life.

References

Abbott, D. M., Jr. (2020). Natural resources and sustainability: Geoethics fundamentals and reality. *The Professional Geologist, 57*(4), 19–25.

Amezaga, J., Balvin, D., Abanto, C., Younger, P. L., & Rötting, T. S. (2007). ERMISA and CAMINAR projects: research on environmental regulation, catchment management and mining impacts in South America. In *Water in Mining Environments. Mako Edizioni, Cagliari*, (pp. 307–311).

APCSG (2011). *Adjuntía Para La Prevención De Conflictos Sociales y La Gobernabilidad: Reportes*. Retrieved February 5, 2011, from http://www.defensoria.gob.pe/conflictos-sociales-reportes.php.

Arellano, J. (2010). *Local politics, conflict and development in Peruvian mining regions* (Doctoral dissertation, Institute of Development Studies, University of Sussex).

Boon, J. (2020). Relationships and the course of social events during mineral exploration: An applied sociology approach. *Springer Nature*. https://doi.org/10.1007/978-3-030-37926-1

Careddu, N., Di Capua G., & Siotto, G. (2019). Dimension stone industry should meet the fundamental values of geoethics. *Resources Policy, 63*, 101468. https://doi.org/10.1016/j.resourpol.2019.101468.

Delman, E. M. (2012). *The History of Mining in Cerro de Pasco and Heavy Metal Deposition in Lake Junin Peru*. Doctoral dissertation. Institute of Development Studies, University of Sussex.

Di Capua, G., Arvanitidis, N., Boon, J., & Nurmi, P. A. (2018, April). The white paper on responsible mining. In *EGU General Assembly Conference Abstracts* (p. 4484).

Durand, F. (2016). *Cuando el poder extractivo captura el Estado: lobbies, puertas giratorias y paquetazo ambiental en Perú*. Oxfam.

García, P. A. (2007, 28 Octubre). El síndrome del perro del hortelano. *El Comercio* (p. A4).

Garcia Vasquez, M. J. (2011). *The role of employee capacity building in reducing mining company-community conflicts in Peru* (Doctoral dissertation, University of British Columbia).

Gardner, E. (2012). Peru battles the golden curse of Madre de Dios. *Nature, 486*, 306–307.

Langeland, A. L. (2015). *Impact of alluvial artisanal and small-scale gold mining in the Madre de Dios River Basin, Peru: total mercury levels in human and farmed fish populations*. Doctoral dissertation *University of Michigan*.

Law 28271 (2004). Law that regulates the environmental liabilities of the mining activity, [online], Congress of the Republic, Peru, July 2, 2004. Retrieved February 12, 2021, from https://www.minam.gob.pe/wp-content/uploads/2017/04/Ley-N%C2%B0-28271.pdf

Li, F. (2013). Relating divergent worlds: Mines, aquifers and sacred mountains in Peru. *Anthropologica*, 399–411.

Li, F. (2016). In defense of water: Modern mining, grassroots movements, and corporate strategies in Peru. *The Journal of Latin American and Caribbean Anthropology, 21*(1), 109–129. ISSN 1935-4932, online ISSN 1935-4940.

Limaye, S. D. (2012). Observing geoethics in mining and in ground-water development: An Indian experience. *Annals of geophysics, 55*(3).

Ministry of Energy and Mines (2020). Updated and Annotated General Mining law 2019. *Supreme Decree N° 014–92-EM*, June 3, 1992, updated and annotated text on April 2018. Retrieved February 10, 2021, from https://www.minem.gob.pe/minem/archivos/file/Mineria/PUBLICACIONES/LGM/LGM%202019.pdf.

Moore, J., Tassé, R., & Jones, C. (2015). *In the national interest? Criminalization of land and environment defenders in the Americas*. MiningWatch Canada.

Nicolau, A. P., Herndon, K., Flores-Anderson, A., & Griffin, R. (2019). A spatial pattern analysis of forest loss in the Madre de Dios region, Peru. *Environmental Research Letters, 14*(12), 124045.

Persson, M. (2009). *Mining activity impact on fluvial sediments in the SW Amazon drainage basin*. University of Gothenburg.

Toledo (2019). Biopolítica, geoética, responsabilidad social y seguridad en la evaluación de la relación Estado, empresas mineras y comunidades. La gran minería en el Perú, el caso del proyecto minero Conga, 1995–2012. Master degree dissertation UNMSM.

Paleontology and Geoethics in Peru: Evolution and Social Practice

Cesar Chacaltana Budiel

1 Introduction

Paleontology is the science that studies fossils in sedimentary and/or epi-clastic rock strata, in all their aspects, modalities and manifestations. This branch of geosciences, due to its transdisciplinary nature and discoveries, has had a great influence throughout history on the world-view of human nature. It has also contributed to the philosophical aspects of thought and the development of scientific morality. Today, research is enriched by the need for geoethical behaviour in light of its discoveries.

From its beginnings in Peru, palaeontology was promoted and linked to the search for mineral resources (Chap. 2 of this book). Throughout the years, its usefulness in geological research applied to the areas of mineral and energy resources was exploited by an academic elite. This stood out for a joint scientific practice that led not only to the creation of institutions, but also to their longevity. The concentration of political-economic power in Lima resulted in the establishment of an oligarchy, to which the scientific elite joined. Cycles of economic crisis contributed to class differences, which led to social unrest over the years. The need to democratize science hence became an ethical commitment for the dissemination of geoscientific knowledge.

Palaeontology has been strengthened with the support of the sectors linked to Education, Culture and Tourism, through exhibitions, the creation of museums and the value of Peruvian geoheritage (Chap. 6 of this book).

Electronic supplementary material The online version of this chapter (https://doi.org/10.1007/978-3-030-86731-7_5) contains supplementary material, which is available to authorized users.

C. Chacaltana Budiel (✉)
Instituto Geológico Minero y Metalúrgico, Lima, Perú
e-mail: cchacaltana@ingemmet.gob.pe

However, in recent decades, this development has also attracted people who exercised scientific practices without accreditation, to the detriment of geoscientist institutions and ethics. Despite the improvement and growth of good research practices in Peru, fraudulent intervention of those who seek to profit from palaeontological research has created division and contradictions in research practices and behaviour. This led to the locking up of the study of Peruvian fossils, under the pretext of the patrimonial protection of "Palaeontological Assets".

These practices serve as the basis for reflecting on the need for training in geoethics, which will assist in denouncing and intervening to prevent these types of actions which harm the socially responsible and sustainable palaeontological research in Peru.

2   Beginnings of Palaeontological Studies

Due to its dynamic geology, Peru has an exceptional palaeontological wealth, with fossil records of different ages and varied nature. They have been found from coastal deserts, in the Andes mountain range and even the plains of the Amazon region. Peru's potential has been recognized since its first discoveries in the "New World", where often anthropocentric views of fossil origins prevailed. The first interpretations of fossils in pre-Columbian America were linked to the stories of diseases or to legends and mythical tales that reflect the spiritual universe of its inhabitants (León, 2016). For instance, during the viceroyalty of Peru, which included territories from Panama to Chile, the presence of fossils was recorded by Cieza de León (1553). The historian described the fossils' discoveries at the tip of the Santa Elena peninsula: "giants came to the coast of the tip of Santa Elena". According to the chronicles, these giants were killed by an angel who with his sword hit and killed them all for having sinned by sodomy: "this is what is said about the giants, which we believe happened because very large bones were found in this part. And I have heard Spaniards who have seen a piece of tooth, which they judged, if it were whole it would weigh more than half a pound of meat" This means that the interpretations in the colonial era fell outside scientific knowledge and were linked to the idealistic conception of monotheistic Christian scripture, widespread in Europe and officially ingrained in the European system. As proof of this, the Quaternary elephant skulls found in Sicily (Italy) were attributed to mythological giants (Fig. 1). During the seventeenth century, the information that Europe had about Peru was very limited, however, there was always a great effort to study the discovered lands, especially by Renaissance cosmographers. This was one of the causes that motivated the organization of scientific exploration trips on this side of the world.

The greatest discovery of the eighteenth century about South America corresponds to the naturalist and geodesist Charles Marie De La Condamine, on a trip promoted by the King of France to measure the terrestrial degrees (Macera, 1976). This expedition was organized in 1735 by the French Academia of Sciences, which was joined by a Spanish mission formed by the mathematician and astronomer Jorge Juan y Santacilia

Fig. 1 Quaternary elephant skulls (*Elephas falconeri*) without tusks, from the Natural History Museum of Verona- Italy, mistaken by Greeks of the time of Homer as Polyphemus, as seen in the marble head of Polyphemus, first century or II d. n.e., Museum of Fine Arts, Boston-USA. Photographs taken from https://archaeopassion.tumblr.com/post/141598070020/cyclops-or-elephants

and the naturalists Jorge Juan and Antonio de Ulloa. This enterprise is recognized because "it was discovered obvious relics of the Universal Flood in the high mountain ranges of the Andes of Peru, in innumerable shells" (Sempere & Guarinos, 1789). All in all, this is evidence that geoscientific studies in Peru at the time were marked by the European stamp promoted by Peruvian political and academic elites, mainly geographers, given the Peruvian mineral wealth. The need to create an open economy motivated an intense search for raw materials, which used scientific research in geology and paleontology.

In that time, the contribution of foreign scientists in the pursuit of understanding the divine also unlocked palaeontological discoveries in Peru and Andes. Alexander Von Humboldt in his trip to America between 1799 and 1804 looked for evidence for expressions of the 'divine' in nature. During that trip, Humboldt made the important discovery of a mastodon molar on the slopes of the Imbabura Volcano in Ecuador, studied by Georges Cuvier who named it in his work "*Fossil Ossements, Tome III Mastodontes*" as *Mastodon des Cordiliéres* (Román, 2010).

A great reader of Humboldt was the French scholar Alcide d'Orbigny, who was considered one of the great naturalists of the nineteenth century and made a trip to South America to undertake zoological, botanical, geological and paleontological inventory. Also, noteworthy in those days was the young Peruvian scientist, Don Mariano Eduardo De Rivero y Ustáriz (Fig. 2), who during his residence in Paris

Eduardo de Rivero   Gustav Steinmann   Carlos Lisson

Fig. 2 Influential scientists who contributed to Peruvian Geology and Paleontology; see references in the text. Online Photographs https://es.wikipedia.org/wiki/Mariano_Eduardo_ de_Rivero_y_Ust%C3%A1riz; https://www.dggv.de/en/about-dggv/honours/g-steinmann-medal. html; https://acecig.wixsite.com/geosciences/post/el-padre-de-la-geolog%C3%ADa-peruana-car los-ismael-lisson-beingolea

met Von Humboldt, to whom he dedicated the mineral called "Humboldtine" that he discovered working in mines in England, France, Germany and Spain (Deustua, 2017).

Giovanni Antonio Raimondi dell'Acqua left a very important legacy in his work "Peru: travel itineraries", published between 1874 and 1913. Likewise, the contribution of William More Gabb stands out, who studied the Peruvian fossils, especially those of Raimondi, such as the holotype *Raimondiceras raimondii* (GABB, 1877). Likewise, Johann Heinrich Conrad Gottfried Gustav Steinmann, whose study "Geology of Peru", synthesizes the Paleontology, Stratigraphy and Tectonics of Peru base for the later work of Carlos Lisson (Fig. 2).

To those contributors can be added the palaeontologists Axel Adolph Olsson and Edward Wilber Berry. The former with his palaeontological studies of the Peruvian Northwest, and the latter with research on the Peruvian paleo-flora between 1900 and 1945.

Then came the era of capitalism derived from the industrial development of Europe, when science and geology courses focused on their contribution to mining. In this context, scientists embraced the philosophical trend of positivism in the country (López, 2012), which led to a utilitarian conception of geology that later placed the country as a relevant producer of raw materials worldwide (Chap. 2 of this book). It is in times of economic growth that the creation of institutions also strengthens international ties. Evidence of ongoing scientific and economic drive, coupled with a national vision for progress, led to the creation of the Geological Mining and Metallurgical Institute (INGEMMET) and the Peruvian Geological Society (SGP) arose. These entities have been collaborating since their inception until present day. For example, in 2018 they organized the II International Symposium of Palaeontology of Peru, an event recognized and congratulated by the international palaeontological community (Fig. 3).

At present, Paleontological research is widely developed through the Palaeontology Area of the Department of Regional Geology of INGEMMET, which is the regulatory body of geological research projects in Peru (Chap. 2 of this book). The vertiginous advance of technology and automated information in the new 21st century has brought with it a new style of a paleontological practice widely used by researchers with a geoscientific vocation. Likewise, the increase in the Public

Fig. 3 Institutional collaboration allowed the II International Paleontology Symposium of Peru. The event was organized by INGEMMET and SGP for the community in general. https://app.ing emmet.gob.pe/evento/IIsimposio/paleoperu

Treasury has strengthened government sectors for an adequate administration that provides facilities, in this case, for scientific research.

There are also educational collections in public universities such as the National University of Engineering with its Paleontology cabinet, the San Marcos National University and its Natural History Museum and the Paleontology Institute of the National University of Piura. Private Universities such as the Pontifical Catholic University of Peru and the Ricardo Palma University also contribute to paleontological collections. The first with the Geological Museum "Georg Petersen" and the second has the Natural History Museum "Vera Alleman Haeghebaert".

3   Geoethics and Paleontology

It is necessary to know details about the current situation of the Peruvian paleontological heritage management in Peru to understand the geoethical problems that have arisen in the professional practice in recent times.

3.1   Paleontological Heritage: Necessary Details

Historically, the idea of heritage has been linked to inheritance. The Royal Spanish Academy defines it in the first order as the "property that someone has inherited from their ancestors" and also defines it as the "set of property and own rights acquired by any title", which is why it is used to refer to the property of an individual. As it is seen, heritage is generally defined as a set of elements to which a value is attributed or assigned, generally for offering some type of real or potential benefit or interest. Likewise, the attributed value is the subjective aspect of all assets and must necessarily exist for it to be considered as such, and must be established by the competent specialist according to the discipline by which it is approached.

The patrimonial aspect of palaeontology is a social construction that it is not given by an inventory or catalogue value, in which all fossil evidence necessarily obtains the category of heritage. During the geological and/or palaeontological exploration process, the sum of fossil finds will indicate the type of fossil and/or deposit and the scientific investigation, its heritage value. Based on this information, what is convenient for an adequate level of protection will be established, since not every object scientifically studied automatically becomes part of the heritage. Therefore, as expressed by Díaz-Martínez et al. (2013) "all assets consist of an objective part (elements that make it up) and a subjective part (their value or interest)". In this regard, to identify paleontological heritage, it is necessary to distinguish the objective part that does not change (the fossils) and the subjective part that can change (the heritage value) (Díaz-Martínez, 2011). For this reason, to assess its value, criteria and research methodologies should be applied exclusively within the geological-paleontological area, as they are objects of nature (Fig. 4).

Fig. 4 Natural manifestation of fossils (Left photo) and manifestation of human activity, reflecting their cultural character (Right photo). The objective part does not change, but as the value is entered, it changes because it is subjective and conditioned by cultural factors. (Photographs from the author's archive)

Regarding that, two fundamental types of heritage should be considered: natural heritage (the result of natural processes) and cultural heritage (the result of human activity), the fossils corresponding to natural and non-cultural elements, therefore it would constitute natural heritage. The foregoing is prescribed in international conventions, especially the provisions of the "Convención sobre la Protección del Patrimonio Mundial, Cultural y Natural" approved by the General Conference of UNESCO in Paris in 1972, which in its article 2 indicates that *"For the purposes of this Convention, the following shall be considered natural heritage: natural monuments made up of physical and biological formations, geological and physiographic formations and strictly delimited areas"*. The palaeontological heritage is the fossils forming part of the geological formations.

On the other hand, the General Law of the Cultural Heritage of the Nation (No. 28296, 2004) states in Article II: "A component of the Cultural Heritage of the Nation is understood to be any manifestation of human or immaterial activity that due to its importance, value in paleontological, archaeological, architectural, historical, artistic, military, social, anthropological, traditional, religious, ethnological, scientific, technological or intellectual significance, is expressly declared as such or on which there is a legal presumption of being. Said assets have the status of public property or private, with the limitations established by this Law". According to this definition, fossils, which are not associated with the manifestation of human activity, should not be governed by this law.

It is worth emphasising that, Peru does not have yet national legislation on fossils, whose conceptualization as paleontological heritage has been defined, characterized or declared (Chacaltana, 2018).

3.2 Questionable Cases

Despite the clarity of the previously exposed definitions, in recent years there have been questionable cases that have evidenced and associated with a practice other than scientific activity. This has led to confusion, and is a detriment to paleontological research in Peru. Moreover, it led to the prohibition of the rescue and research of fossil records.

The cases of professional malpractices were highlighted for the first time in the First International Symposium of Paleontology of Peru. They involve pseudoscientists, who despite having been implicated in questionable events, are accepted by a certain sector of the Peruvian scientific community due to apparent institutional support. Some of them had even obtained official funding, which shows the seriousness of the situation in Peru.

In 2016, during the IX Latin American Congress of Paleontology, a traditional event of the Latin American geoscientific community which has always been successful in the host countries, it was strange that the organization of the event had been awarded to a non-specialist whose only accreditation was his association with the Peruvian Institute of Paleovertebrate Studies. It was announced as an exciting and action-packed event (conferences, specialized symposia, courses, etc.), field trips and a tickets raffle for stays to visit the city of Cusco and Machu Picchu citadel, free of charge. The Ministry of Culture and the Institute of Paleontology of the National University of Piura participated in the organization as well. However, none of the commitments for the event were fulfilled, so it was the subject of journalistic complaints on social networks. Participants requested an investigation in this regard. Indeed, there were many irregularities. Those who participated in the event reported their outrage to feel victims of a scam by the organizers of the event (Gutiérrez-Marco et al., 2016, 2017). Being an international event, the irregularities crossed borders and placed Peru in a level of corruption never before seen at an academic level. The International Society of Vertebrate Paleontology requested the Ministry of Culture of Peru to investigate the case for the damage caused to its associates who, trusting in the organization of the event, participated. The Peruvian Section of the International Association for Promoting Geoethics (IAPG) and the Paleontological section of SGP made a joint communication in this regard to promote a public statement from the geoscience community (SGP, 2016).

What is revealing is that the people who participated in the organization of the event were the same as in 2012 had participated in a meeting for the creation of a "regulation of Paleontological Interventions" in the Ministry of Culture. The contributions and ideas of that meeting ended up being incorporated into the "regulation of Archaeological Interventions" (Supreme Decree No. 003-2014-MC). From that moment on, Paleontological research in Peru from then is conditional on permits requested from the Ministry of Culture. This has forced its intervention in matters far from its function, because of a misinterpretation of the Law.

Furthermore, Law 28296 modifications assigned to the fossils the category of "movable cultural property" and "palaeontological remains" were incorporated into the cultural heritage, without arguments to support it. This generated controversy when trying to grant legal protection to the Ministry of Education in coordination to the Ministry of Culture, even if it was not their field of action. The Law 28296, currently under the supervision of the Ministry of Culture, considers the palaeontological heritage, but under the connotation that it may be a manifestation of human endeavour. Moreover, D.S. 005-2013-MC, which approved the regulation of the Ministry of Culture, incorporated palaeontological concepts such as the so-called "paleontological sites". Based on this, the Guide for the recognition of paleontological assets was developed (Fig. 5). Therefore, in Peru, it is considered legally that nature is culture and geological outcrops have perimeters as if they were archaeological remains.

Fig. 5 Guide for the recognition of paleontological assets. According to this, nature is culture. Photographs by the author and online https://ilamdocs.org/documento/3395/; https://medium.com/@culturaparalima/huaca-palomino-3bd6d191cc2

4 Discussion

As paleontological heritage was included in the Law 28,296, several problems and questions arose about its treatment, conservation and deposition. In this proposition, archaeologists had legal rights to intervene in fossil finds. Therefore, it is not justified that what existed before the advent of humanity, can be within their jurisdiction.

The situation generated unease among Peruvian geoscientists because the Law 28,296 showed not only a conceptual error, but also an epistemological and scientific one. It is a paradox that today, it is still mistakenly considered that such law should include fossils. Considering them as "movable cultural property" generates ambiguities that impose limitations on paleontological research.

Since 2011, new laws proposed on Peruvian palaeontological heritage consider fossils as "movable cultural property". These projects also establish the Ministry of Culture as the governing body of Palaeontology in Peru. However, it is clear that the origin and formation of fossils are exclusively due to nature, biotic and abiotic factors of the environment and its geological evolution. Furthermore, fossils have been formed and deposited without intervention of the human factor, which is indispensable for their consideration as cultural elements. Hence, the state of confusion about the concept is inexplicable.

In 2020, Law Project 5994/2020-CR GENERAL LAW OF THE PALEONTO-LOGICAL HERITAGE OF THE NATION AND PALEONTOLOGICAL SITES was presented, which regulated palaeontological research in cultural fields. Given this action, IAPG Peru wrote an open public letter on this issue (item "Open Letter on Paleontological Heritage" of this book) to promote the reflection on the protection of fossils for their use in scientific research. The law proposal 5994/2020-CR was amended and renamed as GENERAL LAW OF THE PALEONTOLOG-ICAL HERITAGE OF PERU, because of the arguments presented by INGEMMET, the Natural History Museum-UNMSM, CONCYTEC and the Ministry of Culture. The change includes the designation of INGEMMET as the governing body of Paleontology of Peru.

Although, the law is still in observation, the designation of INGEMMET is relevant being the institution that officially develops systematic research on geology, geohazards and geoheritage at the national level, among other functions (Chap. 2).

It is necessary to protect fossils that are distinguished by their rarity, uniqueness, scarcity, exceptional quality or great scientific significance. An ideal way to do this is by promoting paleontological research and creating fossil collections in universities and public or private museums. Moreover, promote the mandatory coordination of paleontological rescue studies when large civil infrastructure works are planned in any part of the country. The unfortunate handling of the subject by the Peruvian institutions is linked to the fact that Earth Sciences are not adequately included in national educational curricula (Chap. 7 of this book). With this perspective, the use of fossils for scientific purposes has been strongly hindered until now.

Finally, it is worth considering that to achieve the continual progression and the development of science, not everything that is included in a legal framework is correct. The history of mankind shows several examples of the need for correction and updating of erroneous norms. As Toledo (2018) mentions, "laws must respond to the incessant need to optimize coexistence and social activity. They must help to solve problems and not create them".

5 Conclusions

The endeavour to observe and catalogue the scientific conception of the world allowed a great pace of development for the natural sciences during the Enlightenment, which coincided with the colonial investigations of Peru. This is especially true for geology and palaeontology, due to their transdisciplinary nature and the dialectical conception of processes and phenomena. In Peru, the advances were promoted by national and foreign researchers, who still serve as a reference for new generations. From the first expeditions and foreign scientific researchers and their national successors, work has continued with the same concerns, knowledge, enthusiasm for ethical practices. Generally, contributions and collaboration of the universities and overseas institutions from different latitudes have furthered the understanding of Peru's mineral abundance, as well as paleontological treasures, as science has no borders.

However, the reported cases of scientific intrusion, fraudulent activities and manipulation of laws make it clear that there is much to be done to protect and regulate research for the benefit of the public, and to protect sites for future generations. Applying Geoethics can help to avoid new situations of this type. National and international cooperation to respond to risky changes in definitions of heritage and, therefore, protections to sites of significance has allowed the geoscience community in Peru to advocate to amendments to potentially harmful laws. These amendments and corresponding rectifications to the described problems will take place for the sake of good practices in geosciences, as well as contributing to scientific research and the establishment of a geoethical awareness in Peru.

Peruvian academia continues with the promotion, research and development of Paleontology in Peru. There is a trend towards teaching geoethical practice at university level to stimulate and prepare young geoscientists. This recognizes the need for new skills in the modern world, with the perspective of responsible behaviour, as a determining factor for decision-making in the country to benefit present and future Peruvians.

References

Balta, J. (1897). Carabaya fossils. *Bulletin of Mines, Industries and Constructions., 14*(3), 19–21.
Benites-Palomino, A., Bellido-Valverde, D., Olmedo-Romaña, G., Burga-Castillo, M., Soria-Hilares, C., & Aliaga-Castillo, A. (2018). The department of vertebrate paleontology of the

museum of natural history UNMSM: What have we learned in 20 years? In *II International Symposium of Paleontology of Peru. Book of Abstracts* (pp. 176–178).

Caravedo, B. (1941). Luis Carranza (Biographical essay). Lima, Printing House of the Victor Larco Herrera Hospital.

Chacaltana, C. (2018). Peruvian paleontology: Historical determination, legal regulations and management proposal. In *II International Symposium on Peruvian Paleontology. Book of Abstracts* (pp. 168–171)

Cieza de León, P. (1553). The Chronicle of Peru. The lordship of the Incas, first part. Seville: in the house of Martin de Montesdoca (Fundación Biblioteca Ayacucho, 2005. Classic Collection, N° 226, Caracas, Venezuela, 473 pp.

Cueto, M. (1989). Scientific excellence in the periphery. Scientific activities and biomedical research in Peru. 1890–1950. GRADE-CONCYTEC, 230 pp.

Deustua, J. (2017). Society, science and technology: Mariano De Rivero, mining and the Birth of Peru as a Republic, 1820–1850. http://www.scielo.org.pe/pdf/apuntes/v44n80/a02v44n80.pdf.

Díaz-Martínez, E. (2011). Typology of heritage: ¿where does geoheritage fit in? En: A. Blieck, P. Auguste y C. Derycke (eds.), Forum GeoReg, Programme and Abstracts, p. 102.

Díaz-Martínez, E., García, A., & Carcavilla, L. (2013). Fossils are geological elements and paleontological heritage is a type of natural heritage. Cuadernos del Museo Geominero, n° 15. Geological and Mining Institute of Spain, Madrid, pp. 583–589.

Galarza, E. (2004). *The economics of natural resources.* Pacifico University. Research Center.

Gabb, W. M. (1877). Description of a collection of fossils made by Dr. A. Raimondi in Perú. *Journal of the Academy of Natural Sciences of Philadelphia, Serie 2,* (8), 263–336.

Gutiérrez-Marco, J. (2013). *Current Peruvian paleontology in the light of international databases: scientists, amateurs and opportunistic intruders.* Book of Abstracts of the I International Symposium on Paleontology. 143 p.

Gutiérrez-Marco, J., Saá, A., & García-Bellido, D. (2016). Two cases of scientific intrusion in Peruvian paleontology. *Geothemes, 16*(2).

Gutiérrez-Marco, J., Saá, A., García-Bellido, D., Chacaltana, C. (2017). Recent geoethical issues in moroccan and peruvian paleontology. *Annals of Geophysics, 60,* Fast Track 7.

Law 28296. General Law of the Cultural Heritage of the Nation. Accessed 22 July 2004 and regulated on 02 June 2006, Peru.

Lay, V. (2005). 2005 annual report. Mining and Metallurgical Geological Institute. 70 p.

León, M. (2016). The contribution of the New World to the birth of paleontological sciences. *Journal of Humanities, No., 28,* 23–48.

López, J. (2012). History of UNI. Vol. I The founding years (1876–1909). UNI University Press., 337 p.

López-Ocón, L. (2001). The geographical society of Lima and the formation of a national science in Republican Peru. Terra Brasilis (Nova Série) Revista da Rede Brasileira de História da Geografia e Geografía Histórica. 22 p.

Macera, P. (1976). The French image of Peru (16th–19th centuries). National Institute of Culture, Lima, 174 p.

Peruvian Corps of Mining Engineers. (1902). Official documents, Boletín N° 1, 47 p.

Ribeyro, J. (1876). *University annals of Peru* (Vol. IX). San Marcos National University.

Román, J. (2010). New contributions to the knowledge of paleontology in Ecuador. In *Proceedings of the 1st International Seminar-Workshop on Geological, Mining and Metallurgical Heritage. Ecuador, National Institute of Geological, Mining and Metallurgical Research.*

Sempere y Guarinos, J. (1789). *Essay of a Spanish library of the best writers of the reign of Charles III.* vol. VI. Royal Printing, Madrid. https://bivaldi.gva.es/es/catalogo_imagenes/grupo.do?path=1002899 (Consulted on February 23, 2021).

SGP. (2016). *Statement on the complaint of fraud during the IX Latin American Congress of Paleontology organized by NGO and the Ministry of Culture in September 2016.* https://www.sgp.org.pe/pronunciamiento-sobre-la-denuncia-de-fraude-durante-el-ix-congreso-latinoamericano-de-paleontologia-organizado-por-ong-y-el-ministerio-de-cultura-en-setiembre-de-2016/.

SGP. (1925). Minutes of society sessions. Sesión I, p. 5–7. Volume One, 126 p.

Tamayo, J., Salvador, J., Vásquez, A., & Zurita, V. (2017). The Mining industry in Peru: 20 years of contribution to the growth and economic development of the country. Lima, Peru. Osinergmin, 316 p.

Toledo, C. (2018). Epistemological and geoethical reflections on geoscientific work in Paleontology. II International Symposium on Paleontology. INGEMMET, Extended Abstracts, pp. 179–181.

Geoethics and the Promotion of Geoheritage in Peru

Julio Cardenas Manzaneda, Sandra Paula Villacorta Chambi, and Carlos Toledo Gutierrez

1 Introduction

The heritage value assigned to geological features encompasses those intrinsic or culturally important at the global, national, state, regional and local levels. These features provide information or knowledge about the geological evolution of geosites and can be used in research and education in Earth sciences (Brocx, 2008; Brocx & Semeniuk, 2015).

A place is considered to be a geoheritage site when the constituent elements of geodiversity have high scientific value (Brilha, 2015). This is the main characteristic for the designation of a Geopark of the United Nations Organization for Education, Science and Culture (UNESCO), which supports the promotion of Geodiversity. Through the geopark concept, UNESCO is promoting sustainable development of the populations involved (UNESCO, 2010). Therefore, geoethics promotes the concept of geoheritage as a social value, so as the creation of new geoparks.

The term, geodiversity, is becoming more commonly used in the geoscience community but despite its importance, it is still largely unknown by the general population (Vasconcelos et al., 2020). Geodiversity is defined as "the natural variety

Electronic supplementary material The online version of this chapter
(https://doi.org/10.1007/978-3-030-86731-7_6) contains supplementary material, which is available to authorized users.

J. Cardenas Manzaneda (✉)
Instituto de Geociencias y Medio Ambiente, Lima, Peru

S. P. Villacorta Chambi
Charles Darwin University, Darwin, Australia
e-mail: sandra.villacorta.chambi@cdu.edu.au

C. Toledo Gutierrez
Antonio Ruiz de Montoya University, Lima, Peru
e-mail: carlos.toledo@uarm.pe

73

(diversity) of geological (rocks, minerals and fossils), geomorphological (landscapes, processes) soil and hydrological features. It includes their assemblages, structures, systems and contributions to landscapes" (Gray, 2004). Brilha et al. (2018) have reviewed the concept to show how geodiversity is connected with other natural systems and highlight its importance in guaranteeing human wellbeing based on the sustainable use of natural resources (Vasconcelos et al., 2020). Speaking on geoheritage, it includes a set of aspects related to the abiotic environment, which involves the lithological, stratigraphic, mineralogical and tectonic characteristics of an area in addition to its geomorphological, pedological and paleontological features (Silva et al., 2014).

Arguments in support of the scientific, educational and touristic importance of geoheritage are abundant (Brocx & Semeniuk, 2007, 2019; Eder, 2008; López-García, et al., 2011) and highlight the need for the promotion of geoheritage and the establishment of legislation to protect it. Therefore, participation and careful evaluation of geoheritage by geoscience professionals is a fundamental social task of great geoethical importance (Gordon, 2018).

In the light of the recognition of the critical role geoscience plays in the development of society, more permanent and sustained efforts are needed worldwide to increase the participation of geoscientists in education, land planning, natural hazards, prevention and other areas. A geoethical approach, therefore, promotes the proper use and conservation of geosites. This implies the dissemination, promotion and, above all, the embracement of the Cape Town Declaration on Geoethics (Di Capua et al., 2017) as a deontological oath. Furthermore, with recognition of serious shortcomings and deficiencies in geoeducation in Peru (Chap. 7 of this book), the geoscientific community plays an important role in the promotion of geoparks conservation from a scholarly level.

2   What Has Been Done for the Promotion of Geoheritage in Peru?

In Peru, the recent development of national policies aimed at social and environmental responsibility has aroused public interest in issues such as geoconservation. In this context, valuing the recommendations of the United Nations, the geoparks' importance to civil society has begun to be recognized. Currently, governmental geoscientific institutions consider research, enhancement and conservation of geosites in their research programs.

The Peruvian geoscientific institution which strongly supports the promotion of geoheritage is the Geological, Mining and Metallurgical Institute—INGEMMET (Fig. 1). It is a specialized, technical public organization, attached to the Ministry of Energy and Mines of Peru. INGEMMET's work is centred on the investigation of basic geology, subsurface resources, geohazards and the geo-environment (Chap. 2 of this book). INGEMMET has developed several studies for the dissemination and

Fig. 1 Cover of Bulletin N ° 9—Series I: Heritage and Geotourism. Geodiversity and Geological Heritage in the Colca Valley. Mining and Metallurgical Institute of Peru (INGEMMET, 2019)

promotion of geoheritage and, in the same way, promoting the creation of geoparks in Peru. Its publications on the subject are open access and available through their Institutional Repository. However, although the work carried out by this organization supports geoheritage in Peru, most of its actions and budget are primarily focused on promoting the development of the mining sector.

3 Colca and Andagua Volcanoes Geopark

In 2003, the Colca Canyon and Volcanoes Valley area in Arequipa, attracted the interest of the Polish Scientific Expedition to Peru due to its high geodiversity and potential for geotourism (Gałaś et al., 2018). The idea and proposal for creating a geopark in that area was led by INGEMMET, which won the approval of regional and local authorities. In 2005 INGEMMET presented the Colca Canyon and Volcanoes Geopark proposal in the workshop "Geoparks and Geoheritage: Promoting the Geoheritage of Latin America" (Fig. 2), jointly organized by the Global Earth Observation Section of UNESCO and the Geographical Institute of the National Autonomous University of Mexico (UNAM). The presentation of the Peruvian geopark proposal at an international event captured the interest of national stakeholders. This facilitated interest and support in the organization of the 1st Geoparks Symposium by INGEMMET, in association with the Regional Government of Arequipa. The event was held in the city of Arequipa in July 2015 (Fig. 3), nearby to the wonderful landscape of the Colca Canyon, one of the main geological attractions of the region. These activities led to the establishment of a Management Committee tasked with the development of the Global Geopark application to UNESCO that was submitted in 2016. The submission was a collaborative effort of professionals from INGEMMET, the Tourism Commission of the Regional Government of Arequipa and AUTOCOLCA (the entity that currently manages the geopark). At the end of 2015, a competition was held to create the Peruvian geopark logo (Fig. 4).

Another result of the effort in promoting the creation of the first Peruvian geopark was the development of the educational modules for use in the geopark

Fig. 2 Presentation of the geopark project: Colca and Andagua Volcanoes in the workshop "Geoparks and Geoheritage: Promoting the Geoheritage of Latin America" UNAM, Mexico 2015

Fig. 3 Poster of the 1st geoparks symposium in Peru. Taken from INGEMMET (2015)

Fig. 4 Contest Poster and logo of the Colca and Andagua Volcanoes Geopark, winner of the contest for its design, convened in December 2015

by the Regional Educational Authority of Arequipa in coordination with the Non-Governmental Organization SOLARIS. Those modules would be officially included in the first stage of the curricula of Primary Education in the southern region of Peru. In this framework, the 5th Argument and Debate Contest called "Geopark of the Colca and Andagua Volcanoes: Is it a Sustainable Development Proposal for the Arequipa Region?" (Fig. 5) was held in November 2017.

Following the Global Geopark submission, UNESCO tasked the Management Committee with the organization of the IV Latin American and Caribbean Symposium on Geoparks in Arequipa (Fig. 6), which had a resounding success in the Latin American region. An outcome of the symposium was to establish the network of Geoparks of Latin America and the Caribbean (UNESCO, 2017).

After the requirements of the Global Geopark application process were met, efforts were crowned with success on 17 April 2019, when UNESCO officially declared the establishment of the Colca and Andagua Volcanoes Geopark, along with seven other geoparks.

Today, the Colca and Andagua Volcanoes Geopark is promoting the geosites protection system, increasing tourism and contributing to the sustained development of the Arequipa region. The commitment of citizens is vital to maintain its status as a world geoheritage site.

Although, until recently, joint actions for the promotion of geoheritage in Peru were almost non-existent, several initiatives for the promotion of geosciences and the development of geoheritage projects at the local level emerged at the end of 2020. For example, the Arequipa branch of the Association of Engineers of Peru (specifically, its Geological Engineering Chapter) sponsored geoheritage projects in the city of Arequipa. The Chapter organized awareness-raising events on geoheritage during

Fig. 5 Fifth Argumentation and Debate Contest "Geopark Colca and the Andagua Volcanoes": Is it a Sustainable Development Proposal for the Arequipa Region? November 2017. Organized by the Regional Direction of Education, Arequipa

its anniversary week (Fig. 7) and invited geoscience professionals from the southern region of Peru to attend. The event raised awareness about the rich geological heritage of the Arequipa region and triggered the interest of many professionals in geoparks projects. A 'Geophotography Competition' and discussion on 'Challenges and duties of geologists in Peru' were organized as part of the event, which especially captured the interest of students and young professionals in geosciences (Fig. 8).

4 Discussion

The Peruvian territory has rich geological sites that constitute an important national and world heritage reserve. However, to date, there have been a limited number of initiatives that promote geoheritage in the country and minimal investment from authorities. This is mainly due to the isolated work and insufficient support from academic bodies, institutions and geoscientific associations. Additionally, the lack of financial support from the government for the research, preservation and conservation of geosites is slowing progress of developing these sites into potential future UNESCO geoparks of Peru.

To enable the protection of Peruvian geoheritage adequate legislation is required. This deserves the attention of the Peruvian State and full support by the geoscientific community. The Peruvian administration needs authorities with a modern vision of the application of geosciences in societal development. Public participation in the

Fig. 6 Poster of the IV Latin American and Caribbean symposium on geoparks held in May 2017 Arequipa, Perú

Fig. 7 Flyer of the anniversary week of the chapter of geological engineering, geophysics and Mines of the South Macro Region, 2020

Fig. 8 Photo entitled "Natural and Cultural Heritage fused into a single Landscape" taken in the Colca and Andagua Volcanoes Geopark. It achieved 3rd place in the 1st geophotography contest organized as part of the Chapter's anniversary week. Photograph by Dante Ojeda (2019)

promotion of other geosites at the national, regional and local levels would allow their use in geoscience education. Geoscience specialists trained in communication are also required to convey geoscientific knowledge and content in a simple, friendly and non-technical language to improve broad public understanding of geoscience (Lockwood et al., 2009; Tavares & Santos, 2014). The value of geoparks should also be presented and promoted by geoscientists across various types of public media, including social networks.

The current democratic context of Peru facilitates political advocacy for geoscientists. However, authorities could do more to understand the vital relationship between geoscientific research, geoethics and politics. This is evidenced by the lack of specific commissions for these important responsibilities within geoscientific institutions. The biased view of the traditional scientific approach, which does not consider the social factor, does not contribute to generating alternatives for the preservation and conservation of geoparks in the country. Therefore, the participation of geoscientists in politics is necessary. For example, on topics such as the preparation of the participatory budget of regional and local governments (Political Constitution of Peru, 1993). Strategic alliances should no longer be understood as existing only between the State and private sector, but also include society. The union of efforts between State, private institutions and local communities seems a better alternative to interact proactively in the context of the twenty-first century (Prosser, 2019). The commitment and participation of local communities ensure the development of sustainable geopark projects over time.

In this context, it is recommended that, as the initiator of the geoparks' protection project, INGEMMET should administer a fund that facilitates and supports broad engagement for stakeholders involved in geotourism. As a government institution, it should collaborate with the teaching on geosciences of officers, academics, scientists, professionals from the state institutions. Through programs to promote the creation of research groups and with the support of funding, such as that provided by the Peruvian National Council of Science, Technology and Innovation; the institution could expand its scope in the promotion of geoheritage and the creation of new geoparks.

On the other hand, because of the limited funding available for geoheritage projects, universities should be more active in research in this area. In order to ensure that geoheritage is high on the agenda of future geoscientists, university curricula should be updated to include geoethics which provides some guiding principles of best practices for identifying and protecting sites of geoheritage significance.

Furthermore, in the context of the global pandemic Covid-19, migration to virtuality should not be an obstacle to the research, preservation and conservation of national geoparks. The injection of funding and support into the development of geoparks is valuable to regenerate the tourism sector in Peru, post-pandemic. As long as the risk of the pandemic lasts, the dissemination of knowledge on geoheritage and geoturism can be done through webinars, promotional videos and virtual geotours. Those tasks involve the participation of state, private institutions, geoscientists and the community at large. Such dissemination at a national and international level can contribute to the assessment, conservation and preservation of geosites.

According to Gordon (2018), geotourism studies could also benefit from the integration of the theory, conceptual analysis and practice of nature-based tourism and closer collaboration with the relevant social sciences. As well as supporting geoconservation outcomes, geoturism provides economic, cultural, relational and social benefits for both visitors and host communities. The interconnections between geoheritage and the cultural components of the landscape have antecedents in concepts of aesthetics in different cultures. These interconnections provide a range of opportunities for enhancing the geotourist experience and promoting geoconservation and geoeducation. To this, the use of activities that involve aesthetic, emotional experiences and interpretation through different cultural filters encourage the rediscovery of a sense of wonder, both about the geological stories in the landscape and the human interactions. A holistic service framework allows for the planning, management and policy of geopark conservation, as well as the evaluation of its benefits and rewards for visitors and communities based on the values of geoheritage assets. The conservation of natural resources is a social process that has undergone adaptations and changes over the years. Geoparks are part of this process that draws attention to competing values and participatory management of land use (Galas et al., 2017; Larwood, 2017; Prosser, 2019; Reynard & Giusti, 2018).

The creation of the UNESCO Global Geoparks Network for Latin America and the Caribbean constitutes a milestone in the promotion and development of geoheritage projects in the region. Likewise, there are ongoing local geosites that can be integrated into other large-scale projects, allowing their integrated management (Sánchez-Cortez & Simbaña-Tasiguano, 2018). Awarding geopark status, UNESCO contributes to a better understanding of the area's geological, natural, archaeological, cultural and industrial heritage. It guarantees the proper use of the economic resources allocated for the preservation and conservation of the geoparks and also, the creation of innovative local enterprises, small businesses, sustainable tourism activities and also new jobs, while protecting the geo-resources. In this context, institutions such as the International Association for Promoting Geoethics (IAPG) and the International Geoscience Education Organization (IGEO) play an important role in the awakening of the international geoscientific community to take on the challenge of promoting the geological heritage and the improvement of geosciences education at large scale.

5   Conclusion

In summary, the promotion of geoheritage in Peru began in 2005 and geoethics plays an important role in making the initiative sustainable and in raising awareness about its protection by society.

As it has been seen in this and previous chapters, the appropriate way to protect geoheritage is through coordinated actions that include the participation of geoscientists. They should also be involved at the political level, as this could be the only possibility to promote the development of new laws, including those to protect geoparks. This task requires an ongoing effort, of which geoscientific institutions must be part and have specific functions. For example, the formation of a national and international geoheritage education commission.

Moreover, since geoparks are laboratories for geoscience education, the Peruvian Government can take this opportunity to update the academic curricula on this subject. Geoheritage and geosciences should be disseminated scientifically, but in a simple and community-oriented language. In this way, citizens, especially those in communities around a relevant geoheritage area (geopark), could learn about the geological history of that geosite and make this knowledge part of their identity.

Finally, as experience shows, only what is understood is appreciated, and so it is essential to disseminate the importance of geoparks to generate commitment and civic responsibility in the population. Nothing is guaranteed in the preservation and conservation of geoparks in Peru without a commitment from the citizens.

References

Brilha, J. (2018). Geoheritage: Inventories and evaluation. In E. Reynard & J. Brilha (Eds.), *Geoheritage: Assessment, protection and management* (pp. 69–85). Elsevier.

Brilha. (2015). Inventory and quantitative assessment of geosites and geodiversity sites: A review.

Brocx, M., & Semeniuk, V. (2019). The '8Gs'—A blueprint for geoheritage, geoconservation, geo-education and geotourism. *Australian Journal of Earth Sciences, 66*(6), 803–821.

Brocx, M., & Semeniuk, V. (2015). *Determining geoheritage values.* Springer.

Brocx, M. (2008). *Geoheritage: From global perspectives to local principles for conservation and planning.* Perth, WA: Western Australian Museum. http://www.museum.wa.gov.au/oursites/perth/shop/newreleases.asp.

Brocx, M., & Semeniuk, V. (2007). Geoheritage and geoconservation-history, definition, scope and scale. *Journal of the Royal Society of Western Australia, 90*(2), 53–87.

Dante Ojeda (2019) colocar en referencias después de la línea 276 del libro, la siguiente referencia: https://institutogema.blogspot.com/.

Di Capua, et al. (2017) Colocar en referencias después de la linea 246 la siguiente referencia: https://www.geoethics.org/ctsg

Eder, W. (2008). Geoparks-promotion of earth sciences through geoheritage conservation, education and tourism. *Journal of Geological Society of India (Online archive from Vol 1 to Vol. 78), 72*(2), 149–154.

Gałaś, A., Paulo, A., Gaidzik, K., Zavala, B., Kalicki, T., Churata, D., ... & Mariño, J. (2018). Geosites and geotouristic attractions proposed for the Project Geopark Colca and Volcanoes of Andagua, Peru. *Geoheritage, 10*(4), 707–729.

Galas, A., Galas, S., Zavala Carrión, B. L., & Churata Quispe, D. (2017). Chances of geotourism development in the Colca and the volcanoes of Andagua Geopark (Peru). *1314–2704.*

Gordon, J. E. (2018). Geoheritage, geotourism and the cultural landscape: Enhancing the visitor experience and promoting geoconservation. *Geosciences, 8*(4), 136.

Gray. (2004). Geodiversity: Valuing and conserving abiotic nature.

INGEMMET. (2015). Colocar en referencias después de la línea 243 del libro, la siguiente referencia: https://repositorio.ingemmet.gob.pe/handle/20.500.12544/2084.

INGEMMET. (2019). Bulletin N ° 9—Series I: Heritage and geotourism of the geological, geodiversity and geological heritage in the Colca Valley. Mining and Metallurgical Institute of Perú.

Larwood, J. (2017). Geodiversity—A cultural template. In Larwood, J., France, S., Mahon, C. (Eds.). (2017). Culturally natural or naturally cultural? Exploring the relationship between nature and culture through World Heritage (pp. 16–19). IUCN National Committee, UK.

Lockwood, M., Davidson, J., Curtis, A., Stratford, E., & Griffith, R. (2009). Multi-level environmental governance: Lessons from Australian natural resource management. *Australian Geographer, 40*, 169–186.

López-García, J. A., Oyarzun, R., Andrés, S. L., & Martínez, J. I. M. (2011). Scientific, educational, and environmental considerations regarding mine sites and geoheritage: A perspective from SE Spain. *Geoheritage, 3*(4), 267–275.

Political Constitution of Peru. (1993). Political Constitution of Peru 1993. Law of the Rights of Participation and Citizen Control N° 26300.

Prosser, C. D. (2019). Communities, quarries and geoheritage—Making the connections. *Geoheritage, 11*(4), 1277–1289.

Reynard, E., & Giusti, C. (2018). The landscape and the cultural value of Geoheritage. In E. Reynard & J. Brilha (Eds.), *Geoheritage assessment, protection, and management* (pp. 147–166). Elsevier.

Sánchez-Cortez, J. L., & Simbaña-Tasiguano, M. (2018). Los geoparques y su implantación en América Latina. *Estudios Geográficos, 79*(285), 445–467. https://doi.org/10.3989/estgeogr.201817

Silva, et al. (2014). *Mapping and analysis of geodiversity indices in the Xingu River Basin*, Amazonia, Brazil.

Tavares, A. O., Santos, P. P. (2014). Re-scaling risk governance using local appraisal and community involvement. *Journal of Risk Research, 17*, 923–949.

UNESCO (2010). Guides to integrate the global geoparks network (GGN).

UNESCO (2017). *IV Simposio Latinoamericano y del Caribe sobre Geoparques en Perú*. Retrieved from http://www.unesco.org/new/es/media-services/single-view/news/iv_latin_american_and_caribbean_symposium_on_geoparks_in_pe/.

Universal Declaration of Human Rights. (UN: 1948).

Vasconcelos, C., Schneider-Voß, S., & Peppoloni, S. (Eds.). (2020). Teaching geoethics. Resources for Higher Education. U.Porto Edições, 207 pp., https://doi.org/10.24840/978-989-746-254-2.

Links of Interest

https://repositorio.ingemmet.gob.pe/handle/20.500.12544/14.
https://www.youtube.com/watch?v=ct8hCcxft18&list=PLI7ifO4dH7S1BTy5Vay0XAI5vEzIX SHdi http://globalgeoparksnetwork.org/?p=377.
https://www.youtube.com/watch?v=YuoYzq_u2G4&feature=emb_logo.
https://repositorio.ingemmet.gob.pe/handle/20.500.12544/2084.

Geoscience Education in Peru

**Roberto Greco, Carlos Toledo Gutierrez,
and Sandra Paula Villacorta Chambi**

1 Introduction

The study of planet Earth involves contributions from many branches of knowledge that are generally defined as Earth science or geosciences. That is why the teaching-learning process related to its understanding is called geoscience (or Earth science) education. This educational branch is important not only for a better understanding of the Earth's natural processes but also to understand the impact of human actions on a global scale, on the use of natural resources and on most economic activities and human needs. Since the publication of the document Earth System Science Overview: A Program For Global Change (NASA, 1986), geosciences teaching has become the accepted method worldwide of developing Earth awareness (Locke et al., 2012). The geoscience taught ranges from school to higher education, as well as including informal education and outreach activities. Geoscientists suggest that the understanding of Earth processes requires an appreciation of the whole complex Earth system.

Geoscience education contributes to the development of specific thinking skills such as retrodictive thinking (Frodeman, 1995), 3D visualization skills and the

Electronic supplementary material The online version of this chapter (https://doi.org/10.1007/978-3-030-86731-7_7) contains supplementary material, which is available to authorized users.

R. Greco (✉)
Campinas University, Campinas, Brasil
e-mail: greco@unicamp.br

C. Toledo Gutierrez
Antonio Ruiz de Montoya University, Lima, Peru
e-mail: carlos.toledo@uarm.pe

S. P. Villacorta Chambi
Charles Darwin University, Darwin, Australia
e-mail: sandra.villacorta-chambi@cdu.edu.au

concept of geological time, which challenge the human common perception of time (King, 2008; Cervato & Frodeman, 2012). As geoscience communication is a multi-disciplinary, complex and dynamic process, it involves educational techniques to apply scientific knowledge in the processes of human and socio-territorial development (Macedo et al., 2007). This reinforces the need for geoscience education to contribute to the management of issues such as the ones covered by the previous chapters of this book.

In this framework, proposing activities that could be effective in the local context, this chapter discusses the geoscience education circumstances in Peru from school to higher education, with a focus on engineering geology as one of the university degrees with more engagement with Earth science. Also covered is the participation of the Peruvian geoscientific community in the national education system, in order to obtain a full and comprehensive picture of the situation in Peru. Bonaccorsi et al., (2020) call this kind of network that involve different actors of the geoscience education as Earth Science Education System.

2 Geoscience Education Circumstances in Peru

According to (Guadalupe et al., 2020), Peru has made a great effort to improve the education system, over the main time that the population has grown by 9 times between 1906 and 2016. Over the same period, the school population increased by 60 times, and now almost a third of Peruvian citizens are enrolled in education. This makes the education system the most wide-ranging and widespread State system in Peru.

Peru is second in Latin America as a country in the number of young people that have concluded secondary school. However, the State is not the only agent of education in Peru; private education grew fast after the nineties, when a new law allowed this development. At the moment, one in four school students attend private school (50% if we consider the main cities of Lima and Arequipa), and three out of four new university students are enrolled in private universities (Guadalupe et al., 2020). It should be noted that the geomorphological and urbanization characteristics of Peru, discussed in the previous chapters, make it challenging to bring education to remote centres.

The Peruvian education system is based on:

- 0–2 year olds: Comprehensive early childhood care programs, initial first cycle.
- 3–5 years old: initial second cycle.
- 5–11 years old: Primary education (6 grades).
- 12–16 years old: Secondary education (5 grades).

These last three stages are compulsory basic education.

There are also some special education programs for children with different abilities, intercultural and bilingual teaching and adult education.

Higher education is offered in three formats:

- University, leading to the academic bachelor degrees (5 years), master's (2 years) or doctorate (3 or more years) and professional certification of the degree, in addition to courses that do not involve an academic degree, such as diploma or specialization programs;
- Non-university leading to professional certification, offered by technological and pedagogical institutes and artistic training schools (3 years); and
- Non-university leading to a first academic degree, offered by professional schools through the legislation approved in October 2016 (Law 30,512) (5 years).

2.1 Primary and Secondary Education

Geoscience is included in the curricular program of primary and secondary education (Peru, 2016a, 2016b) in the subject called "science and technology". The program is organized by competences, and geoscience contents appear in: "explain the physical world". This competence is based in the following criteria:

- Understand and use knowledge about living things, matter and energy, biodiversity, earth and the universe.
- Evaluate the implications of scientific and technological knowledge.

The Earth and the Universe are objects of these standard topics of learning in the Peruvian school curriculum:

- Physical and chemical change in the Earth
- The Earth's origin, its composition, its physical, chemical and biological evolution through fossils
- Flux of energy and matter in the Earth
- Meteorological phenomena
- How the biosphere works
- The Earth's surface relief and Earth's internal activities
- Solar radiation and climate zone
- Earth components and movements.

The curricular program also contains a list of expected skills that students should acquire by the end of each school grade. Tables 1 and 2 show the skills related to Earth Science for primary and secondary school.

In Peruvian high school curricula, science is not separated into different subjects such as physics, chemistry, biology and Earth science as is often the case in other countries. This scenario provides the possibility of integrating them. The curricula also emphasize the importance of students testing hypotheses, collecting data, making analyses, and coming up with conclusions—that is, not only to learn about science but to experience it (Fig. 1). For this to be possible, teacher training in inquiry-based science education, as well as in Earth science issues, is essential.

Table 1 Skills related to Earth Science for primary school, as expected at the end of each grade (Perú, 2016a)

Primary school grade	Skills expected at the end of each primary school grade
First grade	– Describe the characteristic of the soil – Explain the importance of water, air and soil for living beings
Second grade	– Describe that the Earth includes a mass of water, air and solid materials – Describe that the soil is a source of nutrients for many living things
Third grade	– Compare the difference show by the climate over a year and across the difference zones of the Earth's surface
Fourth grade	– Describe the different climatic zone by relating them to the different solar energy that they receive and their relief
Fifth grade	– Explain the dynamic character of the Earth's external structure
Sixth grade	– Make relationships between changes in terrestrial relief and the dynamic internal and external structure of the Earth

Table 2 Skills related to Earth Science for secondary school, as expected at the end of each grade (Perú, 2016b)

Secondary school grade	Skills expected at the end of each secondary school grade
First grade	– Explain how the conditions that make possible life developed on the Earth through the evolution of the universe. Describe how the hydrosphere, atmosphere and lithosphere have changed in the last 4,500 million years
Second grade	– Explain how life and the biosphere depend on the flux of energy and biogeochemical cycles – Explain how the causes of climate change can be mitigated by the use of clean energy to produce electric energy
Third grade	– Explain the Earth's relief with relation to earthquakes, volcanism and rock formation due to the Earth's internal energy
Fourth grade	– Take an ethical position, based on scientific evidence, in relation to paradigmatic events and situations where science and technology are questioned due to their impact on society and the environment
Fifth grade	– Describe how, shortly after the origin of the universe, elementary particles gave rise to H and He, from which, and with the activity of fundamental forces (gravity and nuclear attractive force), the diversity of chemical elements originated that is present on Earth and in the universe – Understanding the ethical, social and environmental implications of scientific knowledge and technologies in the worldview and in the lives of people

Curriculum programs from secondary to fifth grade also introduce ethical reflection, which is worth emphasizing, as it opens the possibility for the introduction of geoethics at secondary level as well.

Specific training in Earth sciences teaching provides the teachers with some specific characteristics of reasoning in the geosciences, such as the retrodictive

Fig. 1 Paleontologist Luz Tejada teaches a course for high school students as part of Miner-Lima2016. In such events, pupils have the opportunity to learn by experiencing geosciences. Photo Sonia Bermudez (2016)

thinking involving scientific 'prediction' of the past. The goal is to discover the causes that created the effect that we can see today, preserved in rock sequences as evidence of past events and past environments. This way of thinking is quite unique and is almost like detective work. It has become a fundamental foundation for students pursuing a career in geosciences. Early exposure to this way of thinking is an original contribution of geosciences to the cognitive development of future citizens.

To prepare teachers for geoscience education, the participation of geoscientists in their training is essential. The Peruvian school curriculum opens the doors to the teaching of geosciences, and that is something remarkable since not all national curricula offer this possibility (Greco & Almberg, 2016) but teacher training is essential to make it a reality in school (Villacorta et al., 2020).

2.2 Higher Education

The university degree with most geoscience content in Peru is the Engineering Geology bachelor degree. The number of geologist engineers is dictated more by the capacity of the universities than by the country's demand for those professionals.

There are no ongoing efforts by Peruvian educational institutions to increase the numbers of students enrolled in this bachelor degree.

There are ten Engineering Geology courses currently being offered, 8 in public universities and 2 in private institutions:

Public Universities

- UNMSM—Universidad Nacional Mayor de San Marcos (National University of San Marcos)
- UNI—Universidad Nacional de Ingeniería (National University of Engineering)
- UNP—Universidad Nacional de Piura (National University of Piura)
- UNSAAC—Universidad Nacional San Antonio Abad del Cusco (San Antonio Abad del Cusco National University)
- UNSA—Universidad Nacional de San Agustín (National University of San Agustín)
- UNAP—Universidad Nacional del Altiplano (National University of the Altiplano)
- UNC—Universidad Nacional de Cajamarca (National University of Cajamarca)
- UNDAC—Universidad Nacional Daniel Alcides Carrión (Daniel Alcides Carrión National University).

Private Universities.

- PUCP—Pontificia Universidad Católica del Perú (Pontifical Catholic University of Peru)
- UPN—Universidad Privada del Norte (Private University of the North).

In addition, the geotechnical engineering bachelor degree is offered by four public Universities: UNJBG—Universidad Nacional Jorge Basadre Grohmann (Jorge Basadre Grohmann National University), UNI, UNAP and UNMSM. The main difference between Engineering Geology and geotechnical engineering courses is the inclusion in the latter of civil engineering subjects such as soil mechanics and testing for construction.

For example, if the elements of the degree in Geology Engineering of the UNMSM are analyzed, it is found that geoscience topics cover about 66.4% of career credits, engineering subjects 22.3% and those of humanities and education only 11.3%.

3 Activities of the Geoscience Community

The institutions dedicated to the training of specialists in geosciences are the universities and some public institutions such as the Geological, Mining and Metallurgical Institute (INGEMMET) the Geophysical Institute of Peru (IGP) and others mentioned in Chap. 2 of this book that have pre-professional internship programs, professional training and thesis support. INGEMMET today involves the largest number of Peruvian geologists dedicated to research and education (Machare, 2018).

4 Approach to the Introduction of Geoethics in the Geosciences Teaching at the University Level

The importance of including the ethical dimensions of geoscience in higher education teaching has been stressed in recent years (Bobrowsky et al., 2017, Vasconcelos et al., 2020; Peppoloni & Di Capua, 2017). They point out that it should not be left to the sense of individual responsibility, but collective. Vasconcelos et al. (2020) also emphasize the importance of introducing early-career geoscientists to geoethical thinking, as it highlights the values behind each professional action.

Two elements are clearly relevant in the development of undergraduate students in geosciences. The first is reflection on the advantages of having a conceptual, categorical and theoretical domain, related to geological knowledge. Teaching activity at this level must be accompanied by metacognition exercises (promoting reflection and awareness of the student on their learning). It allows the student to deepen their knowledge of geology. The second reflection is an epistemological perspective (theory of science), on the nature of geological knowledge. This perspective follows the development of geological thinking, that is, how a geologist reasons while working. This theoretical-practical perspective constitutes the basis for investigation and elaboration of plausible hypotheses in geosciences.

The recognition that geoscientific research applies technologies based on scientific knowledge, has allowed, for example, the development of useful and practical tools in mineral exploration. Geoscientists not only develop understanding of the use of test equipment, but also the purpose of its design and application in collecting field information. They should understand the scope of the investigation and the limitations of the data obtained by the instruments and the possibility of developing plausible hypotheses, as the information collected allows.

In summary, in developing understanding of the use of technology to obtain a projection of their observations from the surface to the underground, geoscientists should recognize the scope and limitations of the equipment employed. This thought should be developed at undergraduate level, promoting reflection. In this regard, Bunge's proposal (2014) of using a systemic approach in the production of scientific knowledge provides support for geoscience education being based on scientific experimentation.

According to the previous perspective, the construction of geological thinking is proposed, as a systematic, innovative and creative approach. The challenge for geoscientists is to construct plausible explanations and hypotheses in the investigation of an area. It requires the development of local and regional geological visualization. It implies:

- Promoting the conceptual and categorical domain of the theories in the study of the Earth.
- Teaching to "see" in the field, using concepts, categories and theories. It concerns a training about how to contrast and verify the effectiveness of this system in field studies and practices. The proposal lies in the conceptual-categorical and theoretical domain. It means that concepts and theories will be used as "lenses"

to see in the field. The proposal takes up the observation of Charles Darwin, who in the nineteenth century recommended: "To be a good observer, you have to be a good theorist" (Canguihem, 2009).

- Encouraging the formation of geoscientific thinking. Students must learn to see geological phenomena as evidence of the multifactorial and systemic characteristics of the Earth. In this regard, an attitude should be promoted that questions the object of study, allowing learners to ask deep questions from different perspectives. It can enrich the thinking. At this point, the observation of Foucault (2007) should be noted, when he was asked about the method he used to address problems. Not being able to ascribe his method to a general classification, he described his research work metaphorically as "the movement of a crab". He explained that his approach to the problem was not directly from one direction, but involved going around like a "crab moving". This interesting perspective broadens perspectives and the possibilities of elaborating original, creative and innovative hypotheses to explain and reconstruct geological processes.

Therefore, methodological strategies should be applied in higher education of geosciences that allow:

- Conceptual and categorical domains of research theories and methods. These will guarantee the visualization of the student during their fieldwork. In addition, it will promote metacognition (knowledge from questions that demonstrate the learning process). The conceptual, categorical and theoretical domain will allow the student and future professional in geosciences to recognize the role they must assume in the stages of research and in exercising their profession. The promotion of geoethical reflection remains a task requiring development.
- The promotion of epistemological reflection and the production of geoscientific knowledge. This allows the development of rational, theoretical-practical thinking (Mosterin, 2010). This contributes to geoscientific reasoning competently in the practice of the profession.

An epistemological and conceptual reflection is part of a methodical didactic strategy to promote critical thinking in geosciences, in the process of training the undergraduate student in geosciences. Through this reflection, basic concepts are presented, which can provide a basis for the student's awareness of the several characteristics of geological work. This includes the different branches in which geoscientists work in the exercise of their profession and research. According to Bunge (2014), it is important to introduce the review of concepts and reflections in science and technology.

Science is "a set of rational, verifiable and fallible ideas to elaborate conceptual constructions of the world. Through it, an attempt to improve the natural environment, based on human needs and to create material and cultural assets, has been made. Applied science becomes technology". Science tries to explain the facts through hypotheses, laws and theories (Bunge, 2014). Technology is science used for practical purposes. It can also be understood as a system of well-founded techniques and or the study of them. For instance, technologists apply the scientific method to solve problems of practical interest. They use technology and its procedures.

Technique. According to Bunge (2014), "it is any coherent set of practices or procedures that lead to a predetermined purpose". Since it is a way of characterizing, not only the methods involved should be taken into account, but also the basis of the procedure. The techniques are classified by Bunge (2014) according to their practical and scientific purposes:

- Practical (pragma-technical) when the technique is predominantly utilitarian, that can be:

  - Founded, as a system of rules justified by a scientific discipline.
  - Unfounded, like a system of empirical rules.

- Scientific when the techniques are based on scientific knowledge.

From these definitions, the relevance in the teaching of geosciences is clear, to promote the student's proficiency of these perspectives. The advantage of this approach is that students will be able to understand the work of the geologist in the context of conceptual domains. The approach also allows the development of rational, systematic, non-contradictory hypotheses coherent with reality.

Finally, the geoethical and social repercussions of geoscience activity depend on (Vasconcelos et al., 2020):

- The framework of reference values existing in the contexts in which geoscientists are operating;
- The level of knowledge, scientific and technical preparation and updating of scientists and professionals;
- The degrees of freedom geoscientists have, depending on where they work (industry, academic research, governmental bodies);
- The efficaciousness of their interactions with other professional figures;
- Their perception of the social utility of their actions.

5   Final Considerations and Proposals

Currently, there is no clear roadmap on how to fill the gap between what is proposed in school curricula and how this material is taught in schools in Peru. Geoscience literacy involves the possibility for Peruvian citizens to develop and use their citizenship rights. The Peru school curriculum provides opportunities for geoscience education, but in order to fulfil these opportunities, there is the need for development of a wide teacher training strategy that will require the support of the Peruvian geoscience community. Geoscientists should undertake leadership of organizing this process as a geoethical duty.

University academic staff should specialize not only in the subject of their expertise, but also in teaching (Fig. 2). This should be an important criterion for the recruitment of professors. The experience of the International Geoscience Education

Fig. 2 Dr. Jose Machare, specialized in mining exploration and also in teaching, sharing his knowledge with young-Earth science students. The image is part of the photo collection of the structural geology course organized by IAPG-Peru in 2018. Photo Sonia Bermudez (2018)

Organization, which promotes geoscientific thinking and metacognitive activities during the learning process (King, 2019), provides examples that should be taken into account in Peru and the Latin American region.

Many actors are involved in geoscience education, and synergic action is required to improve the situation in Peruvian society. Hence, there are some recommendations from the reflections of this chapter:

- Encourage the development of geoscientific thinking.
- Introduce basic Earth science courses in university degrees, as is currently happening with ecology, maths, chemistry courses and so forth.
- Include the active leadership of geoscientists in initial and continued teacher training courses.
- Raise geoethical reflection in society.

The implementation of these proposals requires a cultural change in the attitude and behaviour of citizens and synergistic action by authorities and the geoscientific community. The introduction of geoethics in the Peruvian educational system will lay the foundations for its application in civic awareness. It thus contributes to the solution of the problems currently facing Peru.

References

Bobrowsky, P., Cronin, V., Di Capua, G., Kieffer, S., & Peppoloni, S. (2017). The emerging field of Geoethics. In *Scientific Integrity and Ethics: With Applications to the Geosciences* (pp. 175–212). Special Publications 73. Washington, DC: American Geophysical Union; Hoboken, NJ: Wiley. https://doi.org/10.1002/9781119067825.ch11.

Bonaccorsi, E., Occhipinti, S., Borghini, A., & Greco, R. (2020). Student enrolment in geology from a systemic earth science education perspective: An Italian case study. *European Geologist, 50*, 34–38.

Bunge, M. (2014). *Emergence and convergence. Qualitative novelty and the Unity of knowledge* (pp. xiv + 330). Toronto: Toronto University Press, (Toronto Studies of Philosophy). ISBN 13: 978-14-4262-821-2 (paperback).

Canguilhem, G. (2009). *Estudios de historia y de filosofía de las ciencias* (pp. 25–41). Editorial Amorrortu. Argentina.

Cervato, C., & Frodeman, R. (2012).The significance of geologic time: Cultural, educational, and economic frameworks. *Geological and Atmospheric Sciences Publications, 16*.

Foucault, M. (2007). Nacimiento *de la biopolítica* (pp. 32–54). Fondo de Cultura Económica. Argentina

Frodeman, R. L. (1995). Geological reasoning: Geology as an interpretive and historical science. *Geological Society of America Bulletin, 10*, 960–968.

Greco, R., & Almberg, L. (2018). Earth science education: Global perspectives. E-book. Pouso Alegre: IFSULDEMINAS, 2016. 355 p.: il. ISBN: 987-85-67952-14-7.

Guadalupe, C., León, J., Rodríguez, J. S., Vargas, S. (2020). Estado de la educación en el Perú. Análisis y perspectivas de la educación básica. Impresiones y Ediciones Arteta E.I.R.L., Lima, Perú, pp. 276.

King, C. (2008). Geoscience education: an overview. *Studies in Science Education, 44*, 187–222. https://doi.org/10.1080/03057260802264289.

King. (2019). Exploring geoscience across the globe. E-book. 238 p.: International Geoscience Education Organization. ISBN: 978-1-9996264-0-2. http://www.igeoscied.org/teaching-resour ces/geoscience-text-books/.

Law 30512. Law on institutes and schools of higher education and the public career of their teachers. Congress of the Republic, Peru, 31 Oct 2016. https://busquedas.elperuano.pe/normas legales/ley-de-institutos-y-escuelas-de-educacion-superior-y-de-la-c-ley-n-30512-1448564-1/ (Consulted on February 22, 2021).

Locke, S., Libarkin, J., & Chang, C. Y. (2012). Geoscience education and global development. *Journal of Geoscience Education, 60*(3), 199–200.

Macedo, L., Mariño, J., Fidel, L., Luna, R., Quispe, R., Pareja, H., Arguedas, Á., Siu, A., Muñoz, F., & Ampuero, F. (2007). *Method; proceso de difusión, educación, sensibilización y acción frente a los peligros volcánicos del Misti en Alto Selva Alegre* (p. 60). INGEMMET, Special Publication, Lima.

Macedo, L., Villacorta, S., Vasquez, S., Mariño, J., & Di Capua, G. (2014). Geoscientific communication problem with communities for disaster prevention and land planning in Peru. In *Engineering Geology for Society and Territory* (Vol. 7, pp. 81–83). Springer.

Machare. (2018). Organization of geoscience research in Peru and worldwide. Lecture in "Geoscientific Friday" organized by INGEMMET, Accessed 18 May 2018. Retrieved from https://www. youtube.com/watch?v=r3DauJhWsN4.

Mosterin, J. (2010). *Epistemología y racionalidad* (pp. 21–25). Fondo editorial de la Universidad Inca Garcilaso de la Vega. Lima, Perú.

NASA Advisory Council. Earth System Sciences Committee. (1986). *Earth system science overview: A program for global change*. National Aeronautics and Space Administration.

Peppoloni, S., & Di Capua, G. (2012). Geoethics and geological culture: Awareness, responsibility and challenges. *Annals of Geophysics, 55*(3).

Peppoloni, S., & Di Capua, G. (2017). Geoethics: Ethical, social and cultural implications in geosciences. *Annals of Geophysics.*

Perú, Ministerio de Educación. (2016a). Programa curricular de Educación primaria (p. 396).

Perú, Ministerio de Educación. (2016b). Programa curricular de Educación secundaria (p. 259).

Piscoya, J. C. (2016). Geoethics: Importance of democratizing geosciences in Peru. Lecture in "Miércoles Geológicos". *Peruvian Geological Society.* Accessed 03 Apr 16.

Practical Actions. (2017). Peru 2017 Risks, disasters and reconstruction.

Sucapuca. (2017). IAPG-Peru activities: Promoting adequate application of Geosciences. Lecture in "Miércoles Geológicos". *Peruvian Geological Society.* Accessed 23 Aug 17.

Toledo, C. (2014a). *Operaciones intelectuales cognitivas y la pertinencia del desarrollo de la imaginación lógica creativa en la enseñanza de la geología.* Ponencia en el XVII Congreso peruano de geología.

Toledo, C. (2014b) *"De la necesidad de fundamentos epistemológicos, pedagógicos y didácticos en la enseñanza de la geología en el S.XXI".* Ponencia en el XVII Congreso peruano de geología.

Vasconcelos, C., Schneider-Voß, S., & Peppoloni, S. (Eds.). (2020). Teaching geoethics. Resources for higher education. U.Porto Edições, 207 pp., https://doi.org/10.24840/978-989-746-254-2.

Villacorta, S. P., Sellés, J., Greco, R., Oliveira, A. M., Castillo, A. M., & Regalía, D. A. (2020). LAIGEO, the South American chapter from IGEO and its actions to promote the improvement of Geosciences education in Latin America. *Serie Correlación Geológica, 35*(2), 67–76.

Conclusions: Future Perspectives of Geoethics in Peru

Sandra Paula Villacorta Chambi, Carlos Toledo Gutierrez, and Luis Araujo Ramos

1 Introduction

After the exposition of the Peruvian complex situation, it is evident the importance of applying a geoethical approach, not only to reflect on how to protect the geosphere, but also to promote the Peruvian geoscience's community action within in favour of the resolution of the problems facing Peru.

Such actions for the benefit of Peruvians require geoscientists who are on the front line to educate and establish a dialogue with the Peruvian community. In this context, the dissemination of correct and verifiable scientific information has a geoethical importance. Geoethics is a great opportunity for the scientific community to provide formal and substantial value to the commitment of science for the benefit of society (Peppoloni & Di Capua, 2012). According to Vasconcelos et al. (2020), if society is unaware of the importance of the use of geoscientific knowledge there could be two consequences: the marginalization of geoscientists; and the not trusting of science, to adopt preconceived ideas in a non-critical way.

On the other hand, the modern world requires more attractive places for the dissemination of geoscientific information. New media technologies, such as social

Electronic supplementary material The online version of this chapter (https://doi.org/10.1007/978-3-030-86731-7_8) contains supplementary material, which is available to authorized users.

S. P. Villacorta Chambi (✉)
Charles Darwin University, Darwin, Australia
e-mail: sandra.villacorta-chambi@cdu.edu.au

C. Toledo Gutierrez
Antonio Ruiz de Montoya University, Lima, Peru
e-mail: carlos.toledo@uarm.pe

L. Araujo Ramos
Cesar Vallejo University, Lima, Peru

networking sites, could be used by geoscientists effectively to counteract the misinformation that prevails worldwide (Martin & Peppoloni, 2017). This applies both to Peru and also the rest of the world, and also allows the geoscientific community to strengthen its presence in society.

2   Challenges Working Through

How to strengthen this presence in developing countries like Peru is one of the challenges of future geoethics. This involves figuring out how to go beyond the discipline and the Academy to broaden the geoethical discussion beyond borders. The other challenges are associated with promoting its research and application. In the latter, the value of geoethics stands out, especially for improving the quality of life of the citizens and to prepare them for facing the problems of the modern world (Wyss & Peppoloni, 2015).

2.1   Strengthening Geoethical Values in the Country

Overcoming the traditional paradigm of passive geoscientists in today's world requires a change of attitude from a classic twentieth-century professional to a proactive 21st-century professional (Rajendran, 2010). The former is characterized by showing exclusively analytical, specialized, action and institutional approaches; while the latter is adapting to the needs of global sustainable development. Despite this trend of world change, in Peru, the role and possibilities of geoscientists are still limited. Peruvian society is adapting to modern-day circumstances of a globalized world, both economically and culturally. Information and knowledge technologies have also changed radically to allow access and ensure proper receipt, storage and handling of specialized data (Toffler, 1980). Moreover, the development of a democratic context in the last 30 years in Peru is generating a new scenario of state–society relationship. The mechanism for citizen participation and control is established by Law No. 26313, which establishes citizen rights of Legislative Initiative and Participatory Budgeting.

Changing the passive citizen paradigm of today's society

Changing the passive citizen paradigm of today's society is a large scale mission for the geoscientists. They must not only master epistemology but also demonstrate how geological knowledge benefits humanity and that their contribution can be decisive in the construction of a new relationship between man and Nature (Peppoloni & Di Capua, 2012).

To achieve these objectives, the image of geoscientists isolated from society and politics, and absorbed in their scientific work must be changed. The passive and receptive position of the geosciences has been a serious historical mistake and a

Fig. 1 Geologists explain to authorities of the Chosica district, the preventive measures that must be taken in the face of the rising flow of the Rimac River. Photo INGEMMET, 2017

disadvantage for its progress. In the current times, now that democracy has been strengthened, it is necessary to involve geoscientists in the political arena, in the hope that citizens will become independent of their professions (Lacreu, 2012, 2015).

In this regard, geoscience professionals, exercising their civil rights, must put the results of their studies on the agenda as a matter of public interest. They need to be available to provide advice and assistance to authorities when an adequate application of geoscientific knowledge is required. In the case of Peru, for example, a great opportunity exists to instruct the authorities on how to deal with disasters in the country (Fig. 1).

This approach is also very useful for the mining sector, as expressed in the "Cape Town Statement on Geoethics". It involves the expectation that any geoscientist who works in this area will make a responsible contribution to achieve an adequate regulation to minimize mineral exploitation impacts (Di Capua et al., 2018).

Geoscientists should support geoscientific knowledge and citizen participation in the political sphere, which is the area where laws and national budgets are decided. The advantages that these bring are evident: the existence of laws and state budgets that promote geoscientific research, its dissemination and appreciation by citizens, in a democratic context. However, as expressed by Wood (2001), although geoscientists have the opportunity and the obligation to produce a change in society, they first need to understand and change the misperceptions of the citizens toward their work and approach.

In addition, new leaders or advisers of public institutions are needed, ready to try out a new modern administration with strategic alliances to achieve the afore-mentioned objectives. They should be proactive leaders who believe in creating departments or commissions, with trained personnel and innovative ideas, capable of designing successful advocacy strategies. The scope of this task must aim to achieve laws that promote, through an assigned budget, geoscientific research and the simultaneous work of spreading geoscientific knowledge to society. From there, a further step will be to incorporate these topics into basic and higher-level education plans (Vasconcelos et al. 2020). Effort and dedication must be invested in the achievement of these goals so that in a few years, a positive change towards the application of geosciences can be seen for the benefit of society.

2.2 Research on Geoethics

The geoethics initiatives underway in Peru are being developed to enhance its strength as a discipline that promotes human quality in researchers, scientists and professionals in their quest to contribute to society. Further actions should focus on promoting geoethics research and scholarly interchange.

Promoting research and scholarly interchange

To increase the interest in geoethics, the first obstacle to overcome will be to get the university faculties of geoscience to accept the topic as relevant to geosciences' degrees. For this, the dissemination and teaching of this new science are required. First to the university authorities and then to the educators. Highlighting how geoethics is directly linked to the social and environmental responsibility of geoscientists will contribute to this task.

On the other hand, as scholarly interchange is becoming an integral part of education, the strengthening of inter-institutional relations between geoscientific entities to promote it in the field of geoethics, is required. This resource allows academics, professors and students of geosciences to consult, share ideas and propose research collaborations responsibly. The benefits of this activity are:

- Students attend educational centres where they learn new strategies and technologies, and sometimes new cultures.
- Trainees develop the ability to adapt to new conditions (getting outside the box).
- Improvement of the resume of participants in the program.

Research priorities to develop

The priority topics to further geoethical research in Peru are linked with the strategies exposed in item "Strategies to Strengthen the IAPG-Peru Actions" of this book:

- Working with communities at high risk in hazard-prone areas.
- Social and environmental responsibility of geoscientists.
- Correct dissemination and use of studies, data and technical-scientific results.
- Development of mechanisms for the coordination, review and implementation of policies between state geoscientific agencies.
- Improvement of educational curricula on geosciences and training of geoscientists dedicated to education.
- How to raise awareness of negative research practices.
- How to raise awareness about the importance of Geoethical values in professional activity.
- Promotion of its inclusion in the geoscientific community.

2.3 Implementation

To incorporate a geoethical culture in Peru, it is proposed to maintain its promotion in different scenarios and through various media. This is necessary to ensure the constant exchange of ideas and coordinated initiatives between the country's geoscientists and the authorities. This joint work allows more geoscientists, who practice geoethical values, to take the lead in solving the problems facing Peruvian society using a geoethical approach. It shall also allow incorporation of geoethics in national educational policy, the valuation of geoheritage, among other initiatives.

Organization and promotion of geoethics in different scenarios

To continue promoting a geoethical approach in Peru and the Latin-American region, it is needed to maintain the organization of events and other outreach activities related to geoethics. This organizing activity is also an opportunity to attract commitment from other professionals, especially young geoscientists. This has been confirmed, for example, by the organizing of events such as the annual geoscientific exhibition MinerLima (item "MinerLima: Lima's International Mineral Show", Fig. 2), in which several undergraduate student associations are willing to collaborate and be part of event's organization. In this case, the alliances of the geoscientific community with strategic partners from the public and private sectors are very relevant.

With the increasing use of social networks for scientific communication, more widespread adoption of these channels should be used by geoscientists. However, considering that the evolution of communication is happening without an established code of conduct for scientific discourse (Hillygus, 2018) and that virtual media can encourage hostile and combative scientific exchange (Sundar, 2015); it is worth reflecting on using the media responsibly. One option to ensure transparent and rigorous geoscientific dissemination is for these types of channels to be managed by geoscientists in collaboration with communication specialists. This collective effort will raise the visibility of geosciences at all levels.

Fig. 2 Young student visualizing mineral structures during MinerLima2019. Events like this allow the continuous promotion of a geoethical approach while educating the population

Finally, as has already been discussed throughout this book, the participation of geoscientists in the political field is fundamental to promoting the application of geoethical values in the sustainable development of society, at the local, regional and national levels.

Strengthening relationships in the geoscientist community

Effective coordination mechanisms must be put in place, both at the national and local levels, to promote continuous interaction between professionals from different entities in the country's geoscientific community.

Calling for the formation of inter-institutional committees capable of developing plans for communication, dissemination and positioning of geoethical values at the national level would be of great help in achieving that goal.

On the other hand, the creation of undergraduate study groups in geoethics is fundamental. This will ensure the participation of new generations of geoscientists in the growth of this science.

Promoting the protection of geoheritage

As seen in Chap. 6 of this book, geological culture and geoethics play an important role in raising awareness about the protection of geoheritage within society. Through coordinated actions at the national and local level, it is possible to promote the importance of the identification and study of new geosites and their protection. This protection has so far focused on rural environments that involve natural heritage and has been incorporated into educational practices on Sustainability. It constitutes a critical tool in the construction of knowledge and values by rural communities (De Tarso Castro et al., 2021). Working with them is very relevant in guiding new cultural, social and ethical perspectives that contribute to transforming current practices that impact the environment. The links between communities, their place of origin and their memories are great resources, where the cultural and natural heritage is often indivisible: man, art and Nature together constitute the territory (Peppoloni and Di Capua (2012). This principle makes it possible, for instance, to raise the awareness of producers as agents responsible for the recovery, maintenance and evaluation of environmental systems (Penkaitis et al., 2020).

Including geoethics in the curricula

It is essential that the geoscientific community promotes training in geoethics as part of the educational curricula of developing countries such as Peru.

Lack of training and resources often creates a sense of insecurity. Thus, it is not possible to engage more people in the geoethical proposal if strategies have not been created to promote its teaching at a national level.

Included in the curricula of geoscience faculties, geoethics guarantees skilled geoscientists who, on the one hand, understand their responsibility to society and express themselves in terms of plausibility rather than certainty (Frodeman, 2003) and on the other hand, are prepared to deal successfully with the country's current problems. In the case of Peru, one of these problems is corruption, which, fortunately, today is not as widespread at the government level among state officials as it has historically been (Quiroz, 2008).

These principles can better help the Academy to introduce geoethics as a basis for generating a new professional training scenario with the positive result of professionals with a holistic vision (Almeida, 2014).

3 Final Reflections

Throughout this book, the different chapters have demonstrated the importance of geoscience and geoethics to the scientific economic and socio-cultural development of Peru. In this regard, it is important to recognize that geoethical reflection is obviously a necessary competence to be implemented in Peruvian universities and schools. With modern curricula that include geoethics in the professional internship program, the geoethical reflection can be implemented. This is only possible in Peru if the authorities, private companies and geoscience community combine efforts to carry out this task. The aforementioned entities will only assume these objectives when they are duly informed and understand the importance of geoethics and its benefits.

The achievement of this synergy will open up new possibilities and fields of action, which, for now, are not adequately understood. This includes investment security in the development of research and economic projects. The geoethical approach will ensure coherent actions that will contribute to the resolution of the current bias towards geoscientists, guaranteeing at the same time the continuity of projects for the sustainable use of natural resources. This will also allow meeting current political and social demands, for example promoting social and environmental responsibility, efficiently. Regarding this, it is understood that the ideal professionals for achieving this goal will initially be those practising geoethical values. They must work on the geoethical training of the next generation of geoscientists with the support of the state and the private sector. Such training is a crucial component in the availability of a new generation of qualified geoscientists to confront the real world with a geoethical reflection. This will be achieved in a few years if the work which has been done so far continues. The challenge is to train a modern 21st-century geoscientist, competent, not only in the academic room but also within the community. These professionals must be involved in the development of policies and be able to solve problems democratically with rational arguments.

For the geoscientific community to adapt to the modern world of constant change, sustainably over time, it is very important to work closely with state authorities (at local and regional levels). Since these are not permanent in their positions, geoscientists should promote legislative initiatives that allow them to participate in long-term national development action plans and projects by advising the authorities of the day (Fig. 2).

The basis for achieving these objectives is to promote strategic alliances between the IAPG and national and private scientific and educational institutions, in both the

Fig. 3 Volcanologist Jersy Mariño explaining the Ubinas volcano hazard map to the former President of Peru Ollanta Humala. Photo Macedo, 2014

sciences and the humanities, to increase the possibility of achieving a political influence in society and the decision-making level. In this way, the geoethical approach will be presented at the Legislative level, promoting the debate on laws that benefit the geoscientific community and the national development of geoethical values (Fig. 3).

References

Almeida, A. (2014). A Geoética no Currículo e na Formação de Professores. *GEOlogos, 11.* https://repositorio.ipl.pt/bitstream/10400.21/11366/1/A%20Geo%C3%A9tica%20no%20Curr%C3%ADculo%20e%20na%20Forma%C3%A7%C3%A3o%20de%20Professores.pdf.

De Tarso Castro, P., Mansur, K. L., Ruchkys, Ú. A., & Imbernon, R. A. L. (2021). Geoethics and geoconservation: integrated approaches. *Journal of the Geological Survey of Brazil, 4*(SI 1).

Di Capua, G., Arvanitidis, N., Boon, J., & Nurmi, P. A. (2018, April). The White Paper on Responsible Mining. In *EGU General Assembly Conference Abstracts* (p. 4484).

Frodeman, R. (2003). *Geo-logic: Breaking ground between philosophy and the earth sciences.* SUNY Press.

Hillygus, D. S. (2018). Navigating scholarly exchange in today's media environment. *The Journal of Politics, 80*(3), 1064–1068.

Lacreu, H. L. (2015). Geosciences for citizen training. In XIV *Chilean Geological Congress. Proceedings* (pp. 469–472). Accessed 4–8 Oct 2015.

Lacreu, H. L. (2012). Political roots of geological illiteracy. In *XVII Symposium on Teaching Geology, Minutes* (pp. 91–99), Huelva (Spain).

Law No. 29313. Law that modifies Law No. 26,300, Law on Citizen Participation and Control Rights. Congress of the Republic, Peru. 6 Jan 2009. http://www.keneamazon.net/Documents/Publicati ons/Policy-Analysis/II.-Anexos/Anexo-V.2-Leyes/2009/Ley-29313.pdf (Consulted on February 09, 2021).

Martin, F. F., & Peppoloni, S. (2017). Geoethics in science communication: The relationship between media and geoscientists. *Annals of Geophysics, 60.*

Penkaitis, G., Imbernon, R. A. L., Vasconcelos, C. M. S. (2020). Pagamento por Serviços Ambientais (PSA): o papel do conhecimento geocientífico no protagonismo social. *Terræ Didatica, 16,* 1–12. https://doi.org/10.20396/td.v16i0.8659281.

Peppoloni, S., & Di Capua, G. (2012). Geoethics and geological culture: Awareness, responsibility and challenges. *Annals of Geophysics.*

Rajendran, C. P. (2010). Challenges in earth sciences: The 21st century. *Current Science,* 1690–1698.

Sundar, S. S. (2015). *The handbook of the psychology of communication technology* (Vol. 32). Wiley.

Quiroz, A. W. (2008). *Corrupt circles: A history of unbound graft in Peru.* Woodrow Wilson Center Press.

Toffler, A. (1980). *The third wave.* William Morrow.

Vasconcelos Clara, Schneider-Voß Susanne y Peppoloni Silvia (Eds.). (2020). Enseñanza de Geoética. Recursos para la educación superior. U.Porto Edições, 207 pp. https://doi.org/10.24840/ 978-989-746-254-2.

Wyss, M., & Peppoloni, S. (Eds.). (2015). *Geoethics: Ethical challenges and case studies in earth sciences.* Elsevier. https://doi.org/10.1016/C2013-0-09988-4

Wood, W. W. (2001). Misperception: A challenge for geoscience. *Ground Water, 39*(1), 1–1.

Printed in the United States
by Baker & Taylor Publisher Services

Printed in the United States
by Baker & Taylor Publisher Services